THE *SHARK* THAT *WALKS* ON *LAND*

MICHAEL BRIGHT

THE *SHARK* THAT *WALKS* ON *LAND*

AND OTHER STRANGE BUT TRUE TALES OF MYSTERIOUS SEA CREATURES

This paperback edition published in Great Britain in 2020 by
Biteback Publishing Ltd
Westminster Tower
3 Albert Embankment
London SE1 7SP

ISBN: 978-1-78590-524-7

10 9 8 7 6 5 4 3 2 1

A CIP catalogue record for this book is available from the British Library.

Set in Sabon and Neutra Text

Printed and bound in Great Britain by
CPI Group (UK) Ltd, Croydon CR0 4YY

CONTENTS

INTRODUCTION

The sea can be pretty scary. Never turn your back on it, for it will engulf you in seconds. When you jump or dive into those deep, dark waters, do you wonder what's down there, what creature is underneath you, looking up, ready to take an inquisitive bite? Author of *Jaws* Peter Benchley and film director Steven Spielberg recognised that fear, for below the waves is a world we do not see and we do not understand – an alien world where we are the briefest of visitors. Yet the sea is our planet's universal signature. With so much seawater sloshing about its surface, the Earth is often described as the 'blue planet', yet this watery world is largely unexplored and filled with extraordinary 'alien' creatures whose creation would stretch even the most vivid imagination.

In this little book tales of ancient and modern mariners are told alongside stories of sea serpents, sea dragons, mermaids and mermen, and the legendary kraken. But this is more than a showcase for mystery beasts: marine biologists from all over the world – the modern myth-busters – are finding all manner of unusual animals in the depths of the sea, many new to science. With increasingly sophisticated technology and underwater craft to explore the deep sea, the ocean's closely guarded secrets are being exposed. Initiatives such as the Census of Marine Life, in which scientists from eighty nations spent ten years (and the work is ongoing) recording the life of the oceans, have resulted in a haul of strange sea creatures being brought up from the depths, spotted from submersibles deep down in the ocean or found washed up dead on the seashore. Here we celebrate some of these discoveries, as well as the mysteries, with a blend of the unknown and the familiar in a miscellany of anecdotes, facts and figures about the creatures that live in the sea, including sightings of animals for which science has no identity or explanation.

Epaulette shark

The shark of the title is a little gem of a shark ... almost cute: the epaulette shark *Hemiscyllium ocellatum*. You will find it at low tide on Australia's Great Barrier Reef but don't be afraid of being gobbled up by some amphibious beach monster: it's only a metre long at best. It gets its name from the huge black spot fringed in white on each side of its body, just above and behind its pectoral fins, for it is said the spots resemble epaulettes on a military uniform. In reality, they are probably giant eyespots that deter predators. However, its most unusual trait is that it can 'walk' on its paddle-shaped fins. It can also swim, just like other sharks, to escape larger predators, but often as not it prefers to walk. If low tide and midday coincide, when the sun can beat down on the exposed reef, or at night when oxygen levels in pools can drop significantly, the little shark, unconfined by the sea, can leave the water. If its pool becomes uncomfortable it can simply walk to another, and if that doesn't help it has another trick up its sleeve: it shuts down the motor nerves in its brain but leaves its sensory nerves active and alert for danger, which reduces the need for oxygen. It then dilates its blood vessels, which lowers its blood pressure by as much as half, and pumps more blood to the brain and the heart itself. With these adaptations it can survive in water depleted of oxygen for over an hour, just enough time for the tide to come in and the oxygen-rich ocean waters to wash over the reef once more – all in all a smart little shark ... and endearing too!

Epaulette shark

To include the primary sources of all these tales would fill a book in itself, so for those who would like to follow up any stories that you read here, you can find a list of sources on my website: www.michaelbright.co.uk.

EXTRAORDINARY OCEAN EVENTS

Mother Nature never ceases to surprise us and no event more so than those remarkable occasions when huge numbers of sea creatures suddenly appear at the ocean's surface. Some are fairly predictable but others are once in a lifetime events that just happen with no warning at all. Here are just a few.

Most years between May and July, off the east coast of South Africa, thousands (and even millions) of marine creatures are on the move. It is the annual 'sardine run', during which enormous shoals of South African pilchards *Sardinops sagax* follow a tongue of cold water that pushes north-eastwards between the warm Agulhas Current of the Indian Ocean and the mainland coast. The numbers are incredible. Shoals can be 7km (4.4mi) long, 1.5km (0.9mi) wide and 30m (98ft) deep, and wherever they go a vast procession of predators follows. About 18,000 common dolphins *Delphinus capensis* live in the area and they corral the fish into tight, short-lived bait balls, each 10–20m (33–66ft) across. The dolphins dart in to pick them off one by one, but other eyes have been watching and following, too. Bronze whaler sharks

Carcharhinus brachyurus and lesser numbers of other shark species are attracted to the mêlée, while Cape gannets *Morus capensis* dive down from above. On the shore, people brave the sharks, such as the dusky sharks *C. obscurus* that follow the shoals inshore and the bull sharks *C. leucas* that patrol the murky water, and wade in, using any container they can find to scoop out the oil-rich fish. While the phenomenon is well known to the general public, why it occurs is an ecological mystery. Whatever the reason, it has become known as 'the greatest shoal on Earth'.

Red tuna crab

From time to time off the coast of Baja California, pelagic red tuna crabs *Pleuroncodes planipes* (actually a squat lobster rather than a true crab) swarm in huge numbers, turning the sea bright red. They are unusual for crabs or lobsters, which normally crawl along on the seabed, in that they swim in the open sea, feeding at upwellings where the ocean currents bring deep ocean sediments to the surface and life flourishes. They feed on plankton collected on hairs on their legs. However, the swarms of these 8cm (3in.) long characters attract unwelcome attention. Tuna, blue whales and other whales, sharks, seals, sea lions, gulls and even bats eat them, and loggerhead turtles *Caretta caretta* travel 12,000km (7,457mi) across the Pacific from Japan just to feast on them. The wind and currents can have an impact too. Windrows of crabs up to a metre (3.3ft) deep are sometimes cast up on beaches along the California coast.

A regular spectacle in the Sea of Cortez off Baja California is the congregation of mobular rays *Mobula munkiana*, known locally as *tortillas*. These smaller versions of manta rays are about a metre across and, looking down into the clear water, it is possible to see hundreds of them at any one time. The extraordinary thing is that they leap from the water, their dark, wet backs glistening and their whip-like tails trailing behind. They soar into the air, flapping or curling their 'wings', some rising up to 2m (6.6ft) above the surface, and then land back on the sea with a clearly audible slap. Others leap several times in succession, making flips – up to three at a time – but why they breach in this way is not known. They could be shifting parasites, or they could be driving their prey to concentrate it in one place, a kind of cooperative feeding … or maybe it's just fun.

A less regular occurrence is the sight of huge numbers of rays on the high seas. In 1975, two ships sailing through the same patch of ocean about a month apart encountered the same phenomenon. They were off the coast of Peru near the Ecuador border when the crews saw rays of all sizes leaping from the sea. The master of one ship, the SS *Bendoram*, described how his ship 'passed through these rays for twenty minutes steaming at 15 knots and they were visible on both sides of the vessel for a distance of at least 2 nautical miles'. Similarly, the third officer of the MV *Australind* reported how the shoal 'extended for many miles'. The species was not identified, but an expert from London's Natural History Museum offered eagle rays *Myliobatis* spp. as a possible name, for they sometimes gather in large numbers at the surface, especially around the Galápagos Islands off the coast of Ecuador. Like the mobular rays, there is no explanation for this behaviour.

Sea snake

In May 1932 the passengers and crew of a steamer chanced upon another unusual phenomenon in the Strait of Malacca off the Malaysian coast: the surface waters were teeming with millions of sea snakes. The 1.5m (5ft) long snakes were moving in a line about 3m (10ft) wide and an astounding 97km (60mi) long. The cause was unknown, although open-ocean sea snakes tend to drift with the currents and winds, so accumulate in drift lines, a possible explanation. Such a large number has not been seen since, but there have been reports of smaller aggregations. Sixteenth-century Spanish explorers logged them off the coasts of Central America and several thousand have been observed in Panama Bay. Helicopter pilots have also reported seeing small groups off Vietnam and Pakistan.

In the 29 July 1880 edition of the *Galveston News* there was an unexpected report from a Captain J. B. Rodgers. He was aboard his schooner, the *James Andrews*, when he came across huge numbers of large and small green turtles *Chelonia mydas*, all on their backs. They covered an area of sea 16km (10mi) long and 13km (8mi) wide in the Gulf of Mexico, near the Texas–Louisiana border. Captain Rodgers said 'the water was covered with them'. He also saw Spanish mackerel 'leaping high in the air' in all directions as if escaping from something. It was another of the sea's abiding mysteries, for no one has explained the phenomenon to this day.

A couple of years later, on 21 March 1882, the Norwegian barque *Sidon* was on a voyage from the West Indies to Boston when it reported a strange encounter. The ship was sailing over the steep drop-off into deep water between Nantucket and Chesapeake when it found itself surrounded by huge rafts of dead great northern tilefish *Lopholatilus chamaeleonticeps* floating on the surface. The weather was cold and stormy and as far as the eye could see there were dead and dying fish, some up to 1.2m (4ft) long. The ship was moving at about 7 knots and it ploughed through the fish for 64–80km (40–50mi). They were carried by the wave crests in the rough sea and slapped loudly against the side of the ship. When the news got out, at least twelve other vessels putting in to Boston and New York reported the same scenes during March and April that year. The crew of one vessel, the brig *Rachel Cone*, was in roughly the same sector of the north-west Atlantic as the *Sidon*, and the master reported having seen similar numbers of dead fish on 10 March and passed through them for seven hours. The barque *Plymouth* reached New York on the same day as the *Sidon* docked and revealed that it had ploughed through dead fish for 111km (69mi). A report by the US fish commissioner quoted a conservative estimate of 1,438,720,000 dead fish, drifting in an area of sea 274km (170mi) long and 40km (25mi) wide. A submarine volcano eruption was suggested as a cause, along with warm water, cold water, lack of food and poisonous gases. There were no signs of disease. Speculation was rife, but no one at the time or since has been able to account for this strange and unique phenomenon in the north-west Atlantic. A coda to this story is that no fishing boat caught a tilefish for ten years, until the *Grampus* took eight in 1892; thereafter stocks appeared to bounce back.

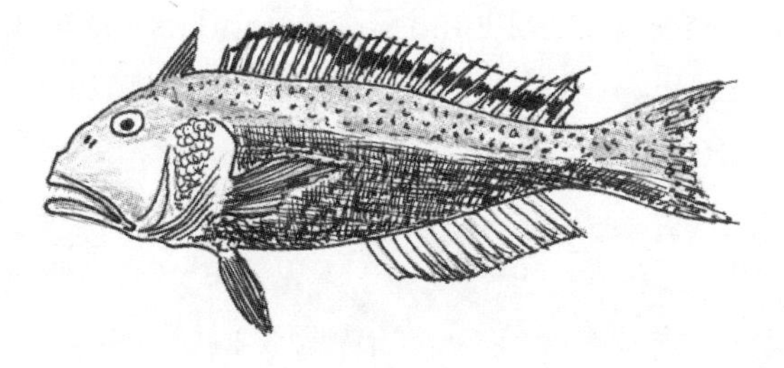

Great northern tilefish

HOLY FLYING ... WHAT!

Flying fish

In May 2008 a flying fish was filmed off Yakushima Island in Japan, where it flew for forty-five seconds, beating the previous record, held since 1920, by three seconds. Flying fish do not fly, as in powered flight, but glide, and there are fish with one pair of wings and others with two. The wings are modified fins and in cross-section they are similar to a bird's wing: curved in such a way as to maximise lift. To become airborne the fish oscillates its tail seventy times a second and speeds along at 70km/h (44mph). On leaving the water, it spreads its wings and tilts them upwards, by which time it is flying through the air, up to a height of 6m (20ft). A typical glide can be 50m (164ft) but if the fish wants to continue, it glides down to the water, wags its tail rapidly, and can gain even greater distances, up to 400m (1,312ft) in a single flight. The record breaker is seen to flap its tail on the sea's surface several times during its flight.

Flying squid can actually fly – it's official! Researchers at Hokkaido University in Japan have discovered that flying squid (Family: Ommastrephidae) have a three-stage flight pattern. First, the squid launch into the air using a high-powered jet of water from their siphon. Second, they spread their fins and fan out their arms to form an anterior and a posterior pair of wings. Webbing between the arms helps to fill in the aerofoil shape of the hind wings. With this configuration they can glide for up to 30m (98ft) at speeds of up to 11.3 metres (37ft) per second – for comparison, Jamaican Olympic sprint champion Usain Bolt reached a top speed of 12.27m (40.26ft) per second at the 2009 Berlin World Championships. They can remain airborne for three seconds, during which time they are able to change their posture and adjust their position in the air. Finally, at the end of the glide, they fold back their fins and arms and dive into the sea. Groups of up to twenty flying squid have been seen flying together. Like flying fish, they are probably trying to escape from predators.

Flying squid

It was the summer of 1894 and Dr Ostrooumoff, director of the Sebastopol biological station, was on a boat excursion along the Crimean coast. It was morning; the sea was calm and the sky an azure blue, when he came across tiny creatures resembling flies hovering above the water. They seemed to sit on the surface and then leap into the air, following a long curved 'flight' back to the sea. He collected a few and took them back to his laboratory. Under the microscope he found them not to be flies at all but marine crustaceans called copepods *Pontellina mediterranea*. He could see that they had feathery appendages that probably lengthened the curve of their fall and kept them momentarily aloft. So, on that momentous morning, crustaceans joined the ranks of animals that could 'fly'. Many years later a Texas University scientist studied how 3mm (0.1in.) long pontellid copepods *Anomalocera ornata* are able to break through the water tension and propel themselves ten to twenty times their body length through the air to escape predatory fish. As a result, these tiny creatures do not have to migrate vertically down to hide in the depths during the day and return to the surface at night, like other zooplankton organisms, but can remain at the surface even in the daytime.

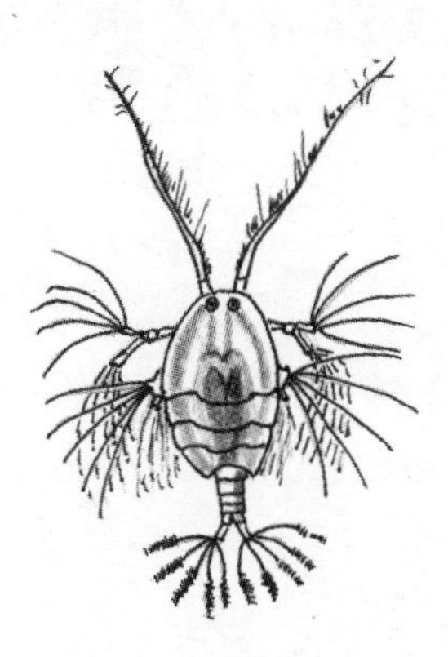

Flying copepod

In December 1912, the American zoologist Dean Worcester was fishing in Bacuit Bay, Palawan, in the Philippines, when he saw a strange low-flying object rising against the wind. He wrote: 'I saw close to my launch what I first mistook for a peculiarly formed flying fish … it was translucent, rose from the water somewhat sharply, and "flew" not more than two or three rods [10–15m or 33–50ft] before dropping into the water again.' However, the more he thought about it, the more he realised that it could not have been a fish. In fact, he later wrote that it looked more like a crayfish or shrimp, about 15–20cm (6–8in.) in length, with 'one or two pairs of flattened legs directed forward and others curving backwards, the legs and the lobes of the tail making the supporting planes'. He saw the same creatures on four more occasions, and was accompanied by a member of the Bureau of Science on one excursion. But what had they seen? I can find nothing since Worcester wrote his most recent note published in 1914.

MERMAIDS AHOY

The true identity of the mermaids described by ancient mariners and made famous by Christopher Columbus on his return from a foray in the Caribbean was probably either a manatee (in the Atlantic and Caribbean) or the dugong (in the Indian and Pacific oceans) … or was it? A flip through the archives of scientific journals and newspapers can turn up some intriguing observations, but they also demonstrate how something from the half-baked world of the paranormal can be so easily accepted as fact.

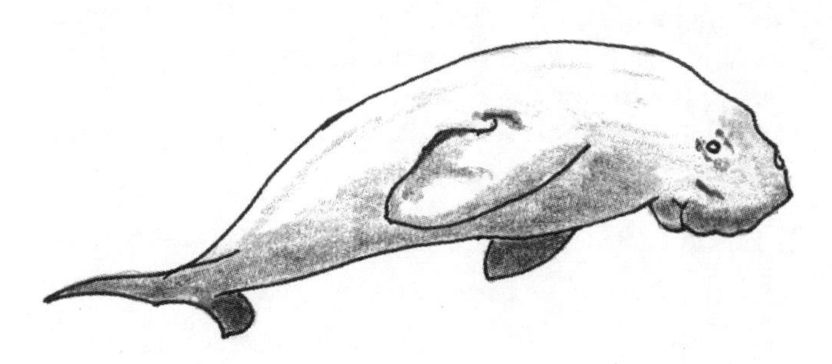

Dugong

In an 1820 edition of the *American Journal of Science and Arts* there is a quite startling extract from a ship's log. The year was 1817 and the ship *Leonides* was on its way from New York to Le Havre, apparently on the 44°N parallel. Its master was Asa Swift.

> At 2 p.m. on the larboard quarter, at a distance of about half the ship's length, [the crew] saw a strange fish. Its lower parts were like a fish; its belly was all white; the top of the back brown, and there was the appearance of short hair as far as the top of the head. From the breast upwards, it had a near resemblance to a human being and looked upon the observers very earnestly; as it was but a short distance from the ship, all the afternoon, we had a good opportunity to observe its motions and shape. No one on board ever saw the like fish, before; all believe it to be a mermaid.
>
> The second mate Mr Stevens, an intelligent young man, told me the face was nearly white and exactly like that of a human person; that its arms were about half as long as his, with hands resembling his own; that it stood erect out of the water about two feet, looking at the ship and sails with great earnestness. It would remain in this attitude, close alongside, ten or fifteen minutes at a time, and then dive and appear on the other side. It remained around them about six hours. Mr Stevens also stated that its hair was black on the head and exactly resembled a man's; that below the arms, it was a perfect fish in form, and that the whole length from head to the tail about five feet.

Mr Elisha Lewis of Newhaven, described as 'a respectable merchant', sent the story to the journal: so far, so good. What Captain Swift and his crew had seen was certainly puzzling, but the journal offers no explanation for the sighting. Could it have been a wayward manatee (after all they are found occasionally as far north as Cape Cod – see also page 276)? Or maybe a seal; they often bob about in the water with their upper quarters clear of the water and have a back end with flippers that resembles superficially that of a fish. Or, was it a sea monkey (see page 251)? Whatever the identity, Captain Swift's mermaid was not the first to appear in a respectable journal.

On 25 May 1809, Miss E. L. Mackay, daughter of the vicar of Reay, a coastal village about 19km (12mi) west of Thurso on the northern tip of Scotland, wrote to the Countess of Caithness about a creature she and her cousin, along with three other people, had seen on the shore at about midday on 12 January. After some preamble, she wrote:

> Our attention was attracted by seeing three people who were on a rock at some distance, showing signs of terror and astonishment at something they saw in the water. On approaching them we distinguished that the object of their wonder was a face resembling the human countenance, which appeared floating on the waves: at that time nothing but the face was visible ... The sea at that time ran very high, and as the waves advanced the Mermaid gently sank under them and afterwards reappeared. The face seemed plump and round, the eyes and nose were small, the former were of a light grey colour, and the mouth was large, and from the shape of the jawbone, which seemed straight, the face looked short ... the forehead, nose and chin were white. The head was exceedingly round, the hair thick and long of a green oily cast, and appeared troublesome to it, the waves generally throwing it down over the face: it seemed to feel the annoyance, and as the waves retreated, with both its hands it frequently threw back the hair, and rubbed its throat ... the throat was slender, smooth and white ... the arms were very long and slender, as were the hands

> and fingers, the latter were not webbed. The arms, one of them at least, was frequently extended over its head as if to frighten a bird that hovered over it, and seemed to distress it much: when that had no effect, it sometimes turned quite round several times successively.

And if you were thinking that Miss Mackay and the others were watching a seal, think again. She went on to say:

> At a little distance we observed a seal. It sometimes laid its hand under its cheek, and in its position floated for some time. We saw nothing like hair or scales on any part of it, indeed the smoothness of the skin particularly caught our attention. The time it was discernible to us was about an hour. The sun was shining clearly at the time. It was distant from us a few yards only. These are the observations made by us during the appearance of the strange phenomenon.

The countess made the contents of the letter public; after all, a respectable vicar's daughter and her close relative are not going to go about pulling the wool over people's eyes. However, its publication prompted another observer to come out of the 'mermaid' closet. In a letter dated 9 June 1809 to a Dr Torrence and published in *The Times* of Friday 8 September 1809, Mr William Munro, a schoolmaster from Thurso, recalled a similar encounter. After berating his reader about the 'general scepticism' surrounding the subject, he went on to describe how about twelve years previously he had been on the beach at the privately owned Sandside Bay, near Reay (the same general area where Miss Mackay had been), when he saw … well, let Mr Munro tell the tale:

My attention was arrested by the appearance of a figure resembling an unclothed human female, sitting upon a rock extending into the sea, and apparently in the action of combing its hair, which flowed around its shoulders, and was of a light brown colour. The resemblance the figure bore to its prototype in all its visible parts was so striking, that had not the rock on which it was sitting been dangerous for bathing, I would have been constrained to [have] regarded it as really an human form, and to an eye it must have appeared as such. The head was covered with hair of the colour as above mentioned, and shaded on the crown, the forehead round, the face plump, the cheeks ruddy, the eyes blue, the mouth and lips of a natural form, resembling those of a man; the teeth I could not discover, as the mouth was shut; the breasts and abdomen, the arms and fingers of the size of a full-grown body of the human species; the fingers, from the action in which the hands were employed, did not appear to be webbed, but as to this I am not positive. It remained on the rock three or four minutes after I observed it, and was exercised during that period in combing its hair, which was long and thick and of which it appeared proud, and then dropped into the sea, which was level with the abdomen, from whence it did not reappear to me. I had a distinct view of its features being at no great distance on an eminence above the rock on which it was sitting, and the sun brightly shining. Immediately before its getting into its natural element it seemed to have observed me, as the eyes were directed to the eminence on which I stood. It may be necessary to remark, that previous to the period I beheld this object, I had heard it frequently reported by several persons, and some of them persons whose veracity I never heard disputed, that they had seen such a phenomenon that I have described, though then, like many others, I was not disposed to credit their testimony on this subject. I can say of a truth, that it was only by seeing the phenomenon, I was perfectly convinced of its existence.

What Mr Munro had seen was certainly unusual, but what follows might offer some clues as to what he saw. What he would not have known is that Miss Mackay's Caithness mermaid might have had a perfectly respectable explanation, revealed by none other than Sir Humphrey Davy (1778–1829). He wrote:

> The mermaid of Caithness was certainly a gentleman who happened to be travelling on that wild shore, and who was seen bathing by some young ladies at so great a distance, that not only genus, but gender was mistaken. I am acquainted with him, and have had the story from his own mouth. He is a young man, fond of geological pursuits; and one day, in the middle of August, having fatigued and heated himself by climbing a rock to examine a particular appearance of granite, he gave his clothes to his Highland guide, who was taking care of his pony, and descended to the sea. The sun was just setting, and he amused himself for some time by swimming from rock to rock, and having unclipped hair and no cap, he sometimes threw aside his locks and wrung the water from them on the rocks. He happened the year after to be at Harrogate, and was sitting at table with two young ladies from Caithness, who were relating to a wondering audience the story of the mermaid they had seen, which had already been published in the newspapers: they described her, as she usually is described by poets, as a beautiful animal, with remarkably fair skin and long green hair. The young gentleman took the liberty, as most of the rest of the company did, to put a few questions to the elder of the two ladies – such as, on what day, and precisely where, this singular phenomenon had appeared. She had noted down not merely the day, but the hour and minute, and produced a map of the place. Our bather referred to his journal, and showed that a human animal was swimming in the very spot at that very time, who had some of the characters ascribed to the mermaid, but who laid no

> claim to others, particularly the green hair and fish's tail; but being rather sallow in the face, was glad to have such testimony to the colour of his body beneath his garments.

So, that takes care of Miss Mackay's 'mermaid' ... or does it? Check the dates. Miss Mackay saw her mermaid on 12 January, whereas Sir Humphrey writes that his bather was in the water in mid-August. So, was there yet another mermaid sighting by two entirely different young ladies at Caithness in the summer of 1809 or has somebody got their dates mixed up and the whole thing is a load of nonsense? Whatever the answer, there is one more possible explanation for the Caithness mermaids.

Scottish historian John M. MacAulay, author of *Seal-Folk and Ocean Paddlers*, suggests people from across the North Sea were involved. These mermaids might well have been 'Sea Saami', indigenous people from Norway's fjords and islands who are thought to have explored Scotland's coasts in their kayaks. They had parkas made from sealskin and their sealskin leg covers could have been mistaken for mermaids' tails. It's a far cry from the romantic image of a mermaid but at least it's a rational explanation.

The sightings, however, go on, this time on the Isle of Man in the Irish Sea. In 1961, the good people of the island, including the Lady Mayoress of Peel, reported seeing mermaids. Commander Roy MacDonald was out fishing when he saw two red-headed mermaids swimming about five miles offshore. 'They were moving at about 12 knots. No human being could swim at such speed.' Unfortunately, they vanished before he could pull up his anchor and investigate. The Lady Mayoress of Peel saw a red-haired mermaid basking off the rocks of Peel Castle, and a local secretary spotted a brunette version perched on a rock to the south of the town. It all prompted the Isle of Man Tourist Board and a Member of the House of Keys to offer a prize of £20,000 for anyone who could capture one alive. I guess that's one way to pull in the tourists, when things are a bit slow.

And if you think they've all gone potty, think again. In May 2012, *Animal Planet* broadcast a spoof documentary with hints of fact. It was titled 'Mermaids: The Body Found' and the National Ocean Service of the USA's National Oceanic and Atmospheric Administration was so inundated with genuine enquiries that they were forced to issue a statement: 'No evidence of aquatic humanoids has ever been found.' So, that's that then.

THE OCEAN'S OLDEST AND YOUNGEST INHABITANTS

In theory, the immortal jellyfish *Turritopsis nutricula* could live, as its name suggests, forever. It can switch back and forth between being a sexually mature jellyfish and a sexually immature polyp, and is the only animal on the planet known to do this. It's a process that in theory can go on indefinitely, but in reality predators and disease put a stop to that.

The oldest continuously growing animal on Earth is thought to be the black coral *Leiopathes glaberrima*, a delicate tree-like coral that has a black skeleton, hence its common name. It grows in deep waters, such as 300m (984ft) down off the Hawaiian Islands. Corals have been found there that are estimated to be at least 4,265 years old. Gold corals *Gerardia* from the same site are also underwater Methuselahs. They were dated at 2,742 years old, and black corals of the genus *Antipatharia*, found in the Gulf of Mexico, are thought to be over 2,000 years old. All these corals grow slowly, no more than a few micrometres per year.

Sponges live long lives. The barrel sponge *Xestospongia*

muta grows slowly but it can reach a height of up to 1.8m (6ft), a giant among sponges. Specimens in the Caribbean have been found to live for at least 2,300 years, and have been nicknamed the 'redwoods of the reef' on account of their age and reddish-grey colour. Similarly, the white or pale yellow Antarctic sponge *Cinachyra antarctica*, which has distinctive tufts of hair-like spicules sticking out of its mainly orange-shaped body, is thought to grow for more than 1,550 years. The spicules deter predatory starfish, enabling the sponge to live to a ripe old age.

The ocean quahog clam or mud clam *Arctica islandica* is a bivalve mollusc found on sandy sea floors in the North Atlantic. It has growth rings in its shell that show its age. One specimen was estimated to be 507 years old, the longest known lifespan for a non-colonial animal. Another was nicknamed Ming after the Chinese dynasty in power when its larva settled on the seabed 405–410 years ago. A 374-year-old clam, kept in an Iceland museum, was found to have fallout from the 1815 eruption of Mt Tambora and other major volcanic eruptions in its shell, each event marked by narrower than normal growth rings.

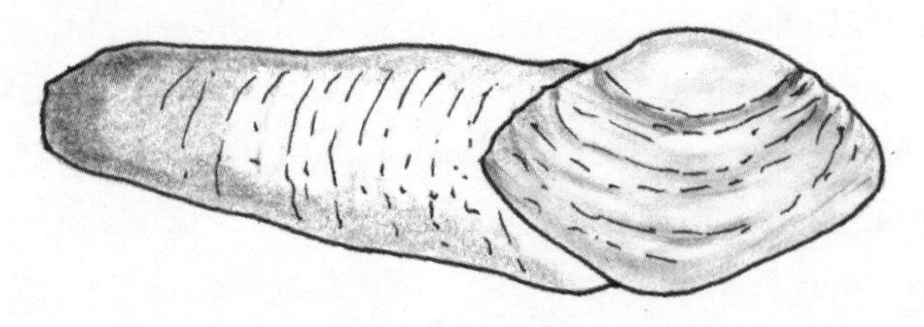

Saltwater geoduck

In fact, several species of shellfish seem to live long lives. Another clam, the giant and edible saltwater geoduck *Panopea generosa*, native to the Pacific coast of North America, is not only the largest burrowing clam in the world (it grows up to a staggering 2m or 6.6ft long), but also one of the oldest. The oldest known specimen was 168 years old, but clams over 100 years old are rare.

The red sea urchin *Strongylocentrotus franciscanus* is a common inhabitant of shallow waters along the Pacific coast of North America. It grazes on seaweeds and, like most sea urchins, has spines to deter predators – a winning combination for it is thought to live up to 200 years old with no signs of age-related diseases. The oldest individuals, each about 19cm (7.5in.) across, live on the coast of British Columbia, between Vancouver Island and the mainland, and their sex drive seems to improve with age. The oldest urchins are the most prolific breeders, producing the largest quantities of sperm and eggs. Maybe this is the reason their sex organs are considered a delicacy in Japan.

The world's oldest mammals are probably bowhead whales *Balaena mysticetus*, which skim zooplankton from close to the surface, and live in Arctic and sub-Arctic waters. The discovery of ancient harpoon heads, some made of stone, buried in the flesh of whales caught more recently by native Alaskan whale hunters, together with evidence of changes in the level of aspartic acid in the whales' eye lens, has led researchers to suggest that these animals can live for more than 200 years. The oldest so far examined was 211 years old. Other whales have shorter lifespans. The oldest known age of a blue whale is 110 years, based on counting waxy laminates, like tree rings, in the earplug of whales. For a fin whale, the figure is 114 years.

Using the same techniques that pinned down the age of bowheads, Danish researchers have examined narwhal eyes from animals that live in the waters off West Greenland, and discovered a female narwhal *Monodon monoceros* to have been 115 years old when she died.

By examining the rings and trace elements in the otoliths (part of the balance organs in the inner ear) of the orange roughy or deep-sea perch *Hoplostethus atlanticus*, researchers have discovered this species of slow-growing, deep-sea fish lives from 125 to 156 years. The brick-red fish is one of the slimeheads, so called because the head is covered in a labyrinth of mucus-filled canals that are part of the lateral line system. In fact, it was always known as the 'slimehead' but commercial fisheries thought the name off-putting so it became a 'roughy'. Why anyone would want to eat the fish is a bit of a puzzle as it is low in omega-3 fatty acids and, because of its longevity, it accumulates large quantities of mercury in its tissues, which if eaten regularly could be damaging to human health. It is found in the Pacific, Atlantic and Indian oceans down to a depth of 1,800m (5,900ft).

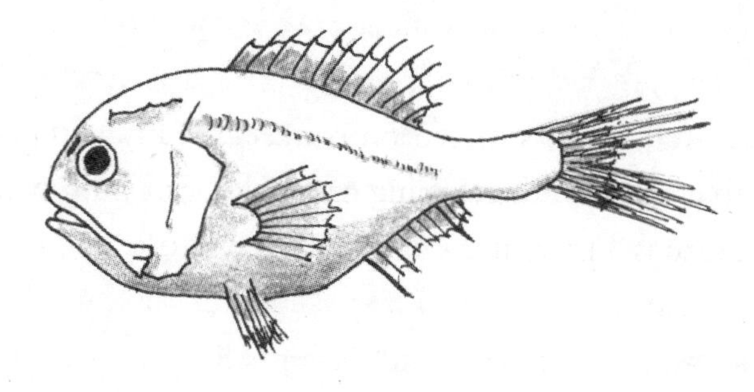

Orange roughy

A lobster *Homarus americanus* caught on the Newfoundland coast and known affectionately as George, was thought to be 140 years old. This would mean it was born in 1869, along with Neville Chamberlain, Mahatma Gandhi and Henri Matisse. It lived in an aquarium in a fish restaurant in New York for a couple of weeks, before public outcry led to its release back into the wild on the Maine coast.

An old orca *Orcinus orca* matriarch, known as J-2 or Granny, is thought to be about 102 years old, as of 2013. She is one of the resident orcas of J pod, which patrols the waters around the San Juan Islands on the Pacific coast of North America. Like most experienced grannies, she is an adept babysitter and is one of the wild orcas to have appeared in the *Free Willy* movies.

At the other end of the age spectrum, the pygmy coral reef goby *Eviota sigillata* from the Indo-Pacific region has the shortest lifespan of any known vertebrate. It lives for just fifty-nine days on average, but grows rapidly. Females produce three clutches of eggs during their short lifespan. The male stands guard and fans the eggs to ensure they receive sufficient oxygen. The hatching larvae float about in the ocean currents for three weeks, before settling on a coral reef where they mature in ten days and live for three and a half weeks, during which time they mate and the whole cycle starts again; a fish that lives fast and dies young.

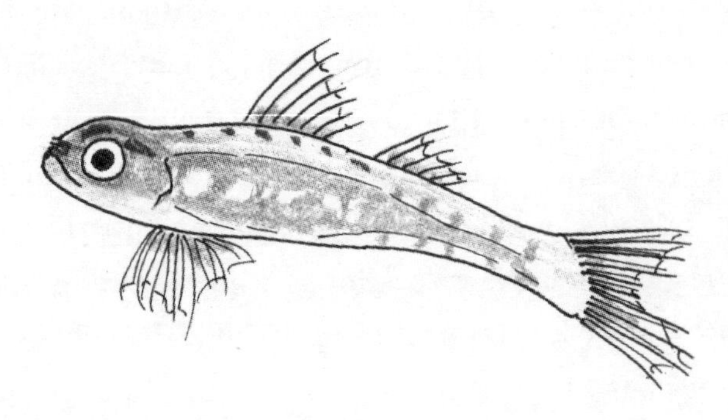

Pygmy coral reef goby

LIVING FOSSILS

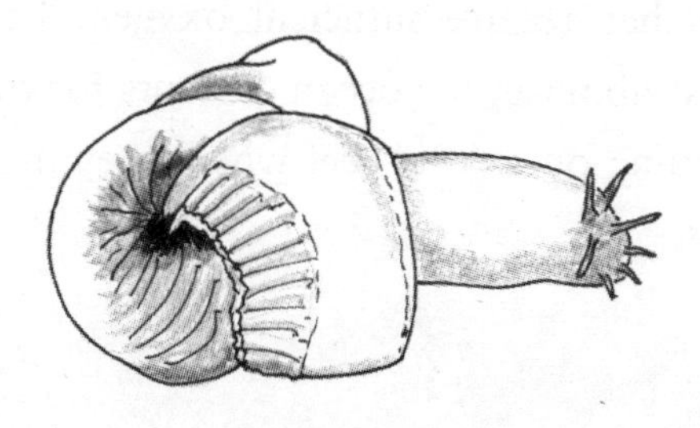

Hagfish

The hagfish (Family: Myxinidae with about seventy-seven species) is probably the slimiest animal on the planet. Glands all over its body can produce enough slime for a hagfish to fill a bucket in minutes. The slime is tougher than most for it contains fibres and so is difficult to remove. The slime affords the hagfish a protective cocoon that can clog the gills of any attacker. The hagfish escapes its own slime by tying itself in a knot and passing the knot along its body to wipe it away. It lives in the deep sea, down to a depth of 1,800m (5,906ft), where often as not it feeds on dead or dying bodies that sink down from above. It bores into the flesh, sometimes using a reverse-knot behaviour (tail-to-head) to help it gain entrance, and then eats it from the inside out, absorbing some nutrients directly through the skin. The hagfish may be eel-like but it is not an eel. It is one of the jawless fish, with a skull but no backbone, and its closest known relative is the lamprey. The fossils of animals similar to modern hagfish can be found in rocks over 300 million years old, making this slimy creature a so-called 'living fossil'.

Coelacanth

The 1.8m (6ft) long coelacanth *Latimeria* is most definitely a living fossil. Fossil remains of similar fish are found in rocks more than 400 million years old and it was thought to have become extinct at the end of the Cretaceous period, about 66 million years ago. However, a specimen was found in a South African fish market in 1938, and since then many more have been seen in deep waters (usually 152–244m or 500–800ft) off the Comoros Islands and at sites across the Indian Ocean to Indonesia. The significance of the fish can be seen in its leg-like lobed fins. Similar fins are seen on ancient fish that lived in the sea 395 million years ago, during the Devonian period, but which were about to crawl out of the water and onto the land, the first tetrapods or four-footed creatures. They also gave rise to the animal's nickname: 'old four legs'. Its feeding behaviour is also unusual. The coelacanth is sometimes seen head down and tail up, using a rostral gland in its snout as a primitive electroreceptor to locate prey, anything from fish to cuttlefish. By day, it shuns the light and rests in underwater caves.

Despite its common name, the pygmy right whale *Caperea marginata*, which lives in the Southern Ocean, is not a right whale, but a living member of an ancient family of whales, known as cetotheres. They were thought to have died out in the late Pliocene epoch about three million years ago, but they are alive and well and swimming in the ocean today. It was not until 2012 that scientists from the University of Otago in New Zealand closely examined the remains of this, the smallest of the baleen whales, from museum collections and made the discovery, so it has also been dubbed a living fossil. An adult pygmy right whale is no more than 6.5m (21ft) long and is thought to feed on shrimp-like krill and other marine crustaceans, but it is so rarely seen that it is still the least studied of all of the baleen whales.

The living animal that comes closest to resembling the legendary sea serpent (see page 285) is the 2m (6.6ft) long, eel-shaped frilled shark *Chlamydoselachus anguineus*. Sometimes it even moves like a serpent, with its head looking as if it is held high on a long neck. In fact, scientists have speculated whether it bends its body and lunges forward like a snake to capture prey in a snake-like mouth that is at the front of its face and not under-slung as in most other sharks. Its most obvious features, though, are the six frilly, naked gill slits on either side of its head. This, together with its strange multi-pointed teeth and primitive appearance, has resulted in this shark joining the ranks of living fossils. Its ancient relatives could well have been swimming in the ocean during the late Jurassic, about 150 million years ago, making it the oldest lineage of living sharks.

Frilled shark

The 3m (10ft) long goblin shark *Mitsukurina owstoni* is definitely a blast from the past; it's a dead ringer for the extinct shark *Scapanorhynchus*, meaning 'spade snout', that swam in Cretaceous seas about 120 million years ago. The living shark is recognised instantly by the long trowel-shaped snout above long, protrusible jaws and a pink-coloured body; but it is rarely seen, for this shark lives in the depths between 200m (656ft) and 1,300m (4,265ft) where little light penetrates. It finds a meal using the many electroreceptors in its long snout. They detect the electricity produced by its prey, such as swim muscles contracting or heart pumping, so it actually has no need to 'see' what it is catching.

Goblin shark with jaws thrust out

The chambered nautilus *Nautilus pompilius* is one of a handful of living nautiloids, a group of marine animals that first appeared about 500 million years ago. Unlike most of their living relatives – squid, octopuses and cuttlefish – the nautiloids have an external, coiled shell filled with gas from which the head, siphon and tentacles protrude. They move by jet propulsion, and the animals can adjust their buoyancy to rise and fall in the water column, moving towards the surface at night to feed on small fish and shrimps beside the outer vertical walls of coral reefs, before sinking back down to about 550m (1,805ft) for the day.

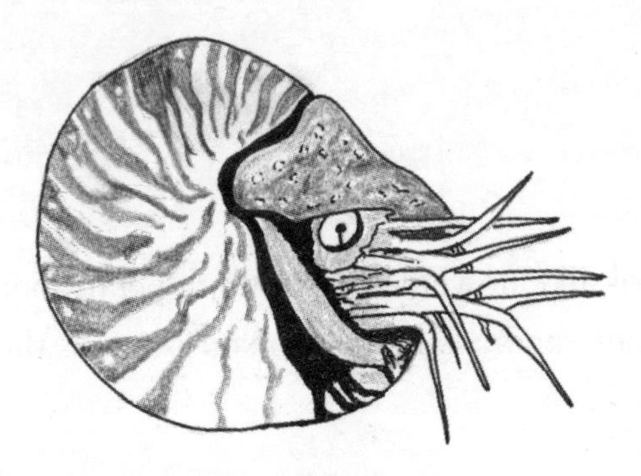

Chambered nautilus

Horseshoe crabs (Family: Limulidae with four living species) look remarkably like extinct trilobites and bear a superficial resemblance to modern crabs and lobsters, but are actually related to spiders and scorpions. Fossils of horseshoe crabs have been found in late Ordovician rocks about 450 million years old. Most of the year they forage on the seabed for marine worms and molluscs, but at breeding time thousands upon thousands of the North Atlantic horseshoe crab *Limulus polyphemus* emerge from the sea and deposit their eggs on beaches on the US east coast. Migrant birds on the East Coast Flyway, arriving from Central and South America, use the beaches as staging posts, the horseshoe crab caviar providing them with fuel to reach their own breeding sites in the Arctic.

Horseshoe crab

They look like flowers with feathery petals, but crinoids are actually animals related to starfish, and their ancestry goes back millions of years. Modern species come in two forms – crinoids with stalks as adults, which are known popularly as 'sea lilies', and 'feather stars' without stalks. They are found at all depths in the sea, from shallow coral reefs to deep-sea slopes. The feathery arms surround a mouth that is on the upper surface at the centre of the 'flower' the opposite way round to starfish, and they filter tiny food particles from the water. The earliest known crinoids appear as fossils in rocks of the Ordovician period about 450 million years ago.

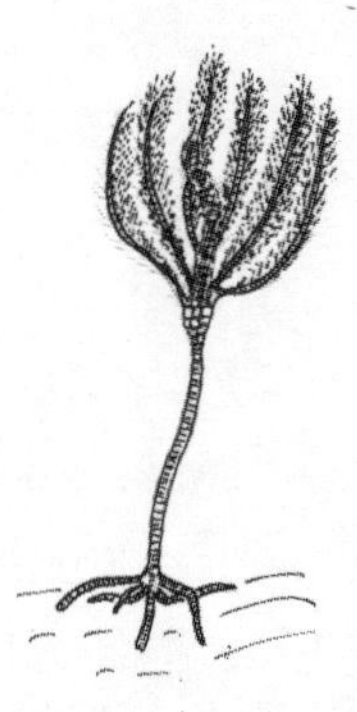

Stalked crinoid

In Hamelin Pool, a shallow lagoon in Western Australia's Shark Bay, the seabed is littered with round, cushion-like structures called stromatolites, the work of cyanobacteria, otherwise known as blue-green algae. These microbes live in shallow water as biofilms in which their cells adhere to each other. The film traps grains of sand and other debris and this gradually builds into a hard mound. Their significance is apparent when one learns that stromatolite fossils have been found in rocks more than 2.7 (and maybe 3.5) billion years old, and therefore represent some of the most ancient records of life on Earth.

SPEEDSTERS

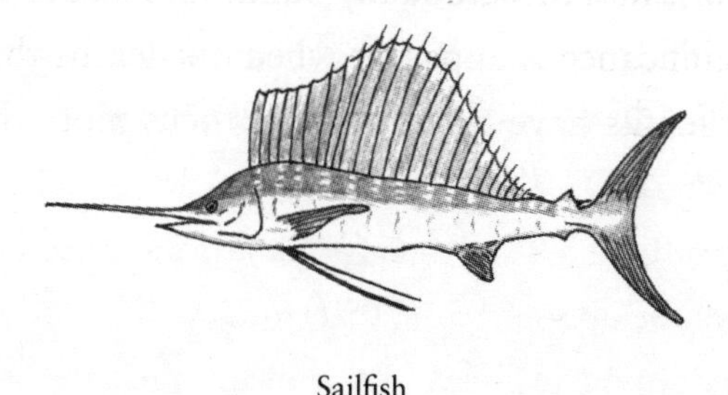

Sailfish

The sailfish *Istiophorus platypterus* is generally recognised by marine biologists as the world's fastest fish. It can swim in short bursts at speeds of up to 109km/h (67.7mph). A sailfish in the Atlantic was clocked making this kind of speed in a series of trials conducted between 1910 and 1925 at Long Key Fishing Camp in Florida. The camp was destroyed with all records in the Labor Day Hurricane of 1935 so reliable documentation does not exist. However, the sailfish is said by former members who remember the trials to have taken out 91m (300ft) of line in three seconds, timed with a stopwatch. It was both swimming and leaping so the timing may not be an accurate swim speed. Nevertheless, it's the most reliable speed trial of a fish achieved so far. The species is recognised by the huge dorsal fin that can be erected, like a sail, in order to frighten and round up fish. The fin folds into a groove when the fish is travelling at speed, and when hunting its body changes from a dull colour to iridescent silvery-blue stripes. Several sailfish might work together to corral a bait ball, flashing their dorsal fins to tighten the ball, before shooting in and whipping their heads from side to side, slashing and maiming the smaller fish to be gobbled up on the next run.

The swordfish *Xiphias gladius* is a large and powerful fish. It can be up to 4.55m (15ft) long, with a body packed with swimming muscles and a brain and eyes kept at a temperature of 10–15°C (18–27°F) above the surrounding seawater. It is not quite as speedy as the sailfish, but it must move at quite a rate and, it seems, has a problem stopping. It has been known to sometimes ram and even sink boats. In 1618 a Dutch mariner was at sea when something hit the ship. When he made port, an inspection of the hull revealed 'a Horne sticking in the Ship' which 'penetrated three planks and turned upwards'. Similarly, the English explorer William Scoresby (1789–1857) wrote about a swordfish beak embedded in the side of ship that docked in Liverpool. The sword had penetrated a sheet of copper, a solid oak timber measuring 19cm (7.5in.), an oak plank 6.4cm (2.5in.) thick and another plank of 5cm (2in.). Had it been withdrawn, Scorseby notes, the ship would have foundered.

Billfish, such as sailfish, swordfish and marlin, can be dangerous, especially when hooked, and there are many instances of them attacking and sinking boats and injuring or even killing crewmembers. In July 2000, a 3m (10ft) long Indo-Pacific blue marlin *Makaira mazara* caught in the Pacific, near Acapulco on the Mexican coast, leaped clear of the water and straight into the boat of Jose Rojas Mayarita. It speared him in the abdomen and the beak came out the other side. He drifted alone in agony for four days until a helicopter on a drug patrol spotted him about 805km (500mi) offshore and another vessel came to the rescue. Unfortunately, the injuries caused an infection from which Mayarita never recovered.

One creature that seems to be able to keep up with billfish and even catch them is the shortfin mako *Isurus oxyrinchus*, the world's fastest shark. It has been clocked at about 50km/h (31mph), with claims of even higher speeds. Accuracy is difficult to assess for the speed of marine creatures is notoriously difficult to establish, but most scientists are agreed that this species is the undisputed champion swimmer in the shark world. Like the swordfish, it is one of the sharks that maintain their swim muscles at a temperature warmer than the surrounding seawater. It also has warm eyes, brain and stomach, all adaptations to be one up on its prey, traits that it also shares with its close relatives the great white and porbeagle sharks.

Shortfin mako

The shortfin mako also holds the speed record for fastest long-distance journey – 2,130km (1,324mi) over thirty-seven days. On average it covered 58km (36mi) per day and all through the relatively dense medium of water. It travels to and from seasonally warm seas, preferring a seawater temperature of 17–20°C (63–68°F), a truly pelagic or open-ocean species. It has a feisty disposition, and is a large and potentially dangerous shark, with forty-two incidents involving it biting humans recorded between 1980 and 2010, together with twenty 'attacks' on boats, although most of those were probably provoked by the shark being hooked on fishing lines and such. Makos have been known to leap straight into boats causing injury to the occupants and damage to the vessel.

Aside from their speed and bad behaviour, these sharks can also be big. The longest reliably measured mako was a female caught off Six-Fours-les-Plages in the south of France in 1973; it was 4.45m (14.6ft) long. In the late 1950s, a giant mako was caught off Marmaris, on the Turkish coast of the south-east Aegean Sea. It was photographed and, as recently as 2010, the images were analysed and a report published by Turkish and Italian scientists. They worked out that the shark was 5.85m (19.2ft) long – with an estimated range of 5.77–6.19m (18.9–20.3ft) – which exceeds the previously recorded maximum size for the species. However, more usually people encounter much smaller sharks, the largest with a length of 3m (10ft).

DEEPEST DIVERS

Sperm whales *Physeter macrocephalus* are thought to be able to dive down to 3,000m (9,843ft) or even further. A sperm whale harpooned off South Africa in 1969 was found to have two newly caught bottom-dwelling sharks in its stomach, and the seabed at the catching site was at least 3,193m (10,476ft) deep. A spotter plane recorded that the whale had been below for a staggering 1 hour 52 minutes. A feat like this affords the sperm whale the status of world's deepest-diving air-breathing animal. However, this dive was exceptional. More usually they head down to 600–1,000m (1,969–3,281ft), where they hunt for squid in a dive that typically lasts about forty-five minutes.

A southern elephant seal *Mirounga leonina* holds the record for the deepest-diving seal or sea lion. An adult male, carrying a depth recorder, went down to at least 2,133m (6,998ft), and another is claimed to have reached 2,388m (7,835ft), but more often they go to 500m (1,640ft) for about half an hour (see page 244).

Cuvier's beaked whale *Ziphius cavirostris* has been tracked to 1,900m (6,234ft) and remained down for eighty-five minutes off the Italian coast; and a Blainville's beaked whale *Mesoplodon densirostris* has been observed to dive to 1,250m (4,101ft) for fifty-seven minutes off the Canary Islands.

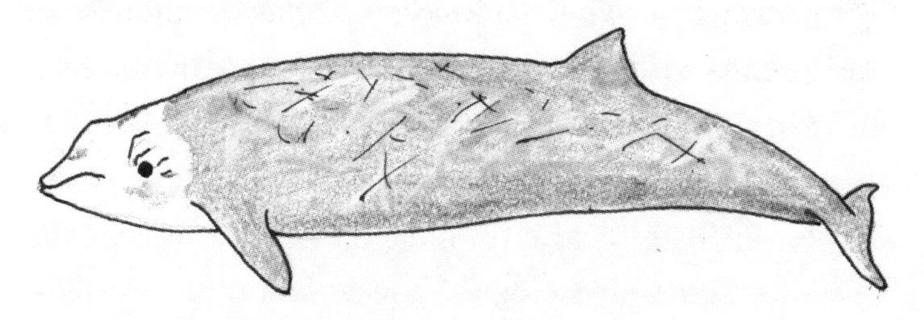

Cuvier's beaked whale

The northern bottlenose whale *Hyperoodon ampullatus* can dive to a depth of 1,453m (4,767ft) and remain under the water for up to seventy minutes, returning to the surface relatively slowly to avoid the 'bends'. Other species of small whales go to such depths, but the northern bottlenose whale is the only one to go deep routinely because it hunts for squid close to the seabed.

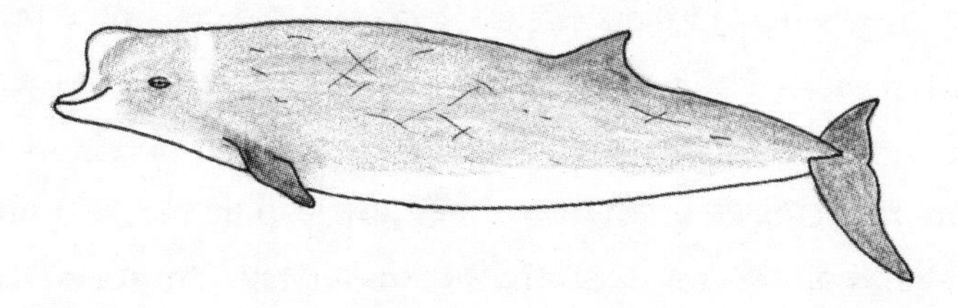

Northern bottlenose whale

Antarctica's Weddell seal *Leptonychotes weddellii* not only has the most southerly distribution of any mammal (apart from us), but it is also an accomplished diver. It goes down to 750m (2,461ft) and can stay underwater for up to eighty minutes.

The emperor penguin *Aptenodytes forsteri* is the largest of the penguins. It has been known to dive to 535m (1,755ft) and remain below for up to 27.6 minutes (but more usually about five to ten minutes), making it the deepest-diving bird with the longest dive time. It achieves this with some physiological tricks. Before it even hits the water, its heart begins to race and it hyperventilates, priming its muscles with oxygen; but when in the water it cuts off the blood supply to its muscles. The heart rate, meanwhile, drops to as little as six beats per minute for the last five minutes of an eighteen-minute dive, pushing its body to the limit, in fact to a point at which a human would pass out. The bird, however, hunts for fish and squid at depth, returning to the surface in a lubricating stream of bubbles. These have been held in the feathers, and their release decreases the density of the water surrounding the penguin and therefore lowers the water resistance dragging against the bird's body so it can rocket back up to the surface and leap through the air to land on the sea ice. Here it takes a deep breath and its heart rate rises to 256 beats per minute (compared to 73 beats per minute at rest), bathing its muscles in oxygen-rich blood so that the bird quickly recovers and is ready to dive again.

Imperial cormorants *Phalacrocorax atriceps* off Patagonia have been filmed diving to 45m (148ft) and have been monitored to 60m (197ft). They stay underwater for about two and a half minutes, 'swimming' down by pumping their wings and feet. Similarly European shags *P. aristotelis* have been monitored down to 45m (148ft), with a dive time of forty-five seconds on average.

Imperial cormorant

The yellow-bellied sea snake *Pelamis platura* dives down to depths of 50m (164ft) in the open ocean and can remain underwater for nearly four hours, according to observations in the Gulf of Panama. It spends as much as 99.9 per cent of its life below the surface. The olive-brown sea snake *Aipysurus laevis* has been tracked down to 68m (223ft); the blue-lipped sea krait *Laticauda laticaudata* and Saint Girons' sea krait *L. saintgironsi* fitted with dive loggers regularly reached 80m (263ft) in a lagoon in New Caledonia and stayed down for 130 minutes; and a reef shallows sea snake *A. duboisii* was caught in a trawl set at 80m (263ft).

The sea otter *Enhydra lutris* on North Pacific coasts dives down to 37m (121ft), and can hold its breath for up to four minutes. It pushes through the water by flexing the rear part of its body, tail and legs with an up-and-down movement, travelling at a top speed of about 9km/h (5.6mph).

By comparison, human free divers can reach 101m (331ft) unassisted, the record holder for 'constant weight without fins' being New Zealander William Trubridge. This is the most challenging of the free-diving categories for the divers must descend and ascend under their own steam.

ZONES OF THE OPEN OCEAN

In order to come to terms with the vastness of the open oceans and the deep sea in particular, scientists have adopted a simple vertical classification system which takes account of the amount of light reaching different depths. It goes like this:

- The sunlight or epipelagic zone extends from the surface to 200m (656ft) where most light and heat penetrates. This results in a wide range of temperatures close to the surface.

- The twilight or mesopelagic zone extends from 200m (656ft) to 1,000m (3,281ft) down. Light from the surface is faint and bioluminescent organisms are common.

- The midnight or bathypelagic zone extends from 1,000m (3,281ft) to 4,000m (13,123ft) down. Little to no light penetrates this far down so the creatures living in this zone make the only light themselves. Many are coloured black or red, making them functionally invisible.

- The abyss or abyssopelagic zone extends from 4,000m (13,123ft) to 6,000m (19,685ft) down. There is no light, the water pressure is immense and the temperature is near to freezing. Much of the ocean floor, known as the abyssal plain, lies in this zone.

- The trenches or hadopelagic zone extends from 6,000m (19,685ft) to 10,911m (35,797ft) down, the latter depth representing the bottom of the world's deepest trench, the Mariana Trench. The water temperature here is almost freezing and the pressure eight tons per square inch, the equivalent of the weight of forty-eight jumbo jets; yet, there is life even in the deepest trenches.

TROUBLE IN THE SUNLIGHT ZONE

Almost all the underwater life with which we are familiar is to be found in the sunlight zone, but in some parts of the world things are not as they seem. Some places are not nice places to live at all, and there are no-go areas too.

On the muddy bottom at depths of 50–200m (164–656ft) off the coasts of Chile and Peru are huge spaghetti-like mats of giant bacteria *Thioploca*, each 2–7cm (0.8–2.8in.) long and visible to the naked eye. They cover an area of 130,000 sq. km (50,193 sq. mi), almost the size of Greece. The microbes live in layers in the deep sea known as 'oxygen minimum zones' or 'shadow zones', where the bacteria live off hydrogen sulphide, a toxic gas that is produced when organic matter drifting down from the surface decomposes in the absence of oxygen. Oxygen levels are so low that few animals can survive here.

The Baltic Sea is home to seven of the world's ten largest dead zones. Oxygen here has been used up by bacteria that have been fed by the decaying remains of blooms of algae that, in turn, are fed by sewage from towns and the fertilisers that run off the surrounding farmland. The algal blooms themselves can be toxic to swimmers, foul beaches and envelop seaweeds, so fewer natural habitats are available for marine life. The end result is that huge areas in this brackish sea are entirely devoid of marine life other than bacteria.

IN THE FAINT GLOW OF THE TWILIGHT ZONE

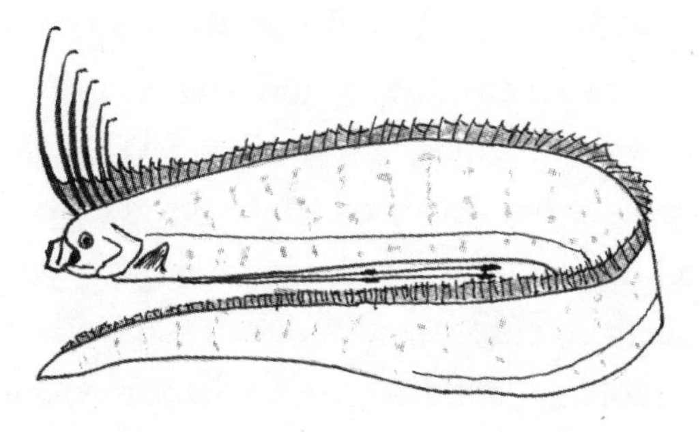

Giant oarfish

The world's longest-living bony fish swims in the twilight zone. It is the giant oarfish or king of the herrings *Regalecus glesne*, with an exceptionally long, smooth, silvery, laterally compressed, ribbon-shaped body, up to 15m (49ft) long according to unconfirmed reports. On the top of its head is a constantly moving cockscomb of long red fin rays with flaps of skin at their tips and along the length of its back is a rippling dorsal fin. These striking fish occasionally come close to the surface and into shallow water, probably when they are sick or dying, and could quite easily be mistaken for a sea serpent (see page 285). In February 2003, British angler Val Fletcher caught one such specimen while fishing from the rocks for mackerel at Skinningrove on the North Yorkshire coast. It was 3.3m (10.8ft) long. The Natural History Museum in London was interested in preserving it but unfortunately Ms Fletcher cut it up for the pot!

Go to the fish market in Funchal on the north-east Atlantic island of Madeira and you are sure to find another twilight zone fish – the black scabbardfish *Aphanopus carbo*, one of the cutlassfish family. It is eel-like, laterally compressed like the toothless oarfish, but unlike them it possesses a mouth filled with seriously sharp, fang-like teeth. It is coloured an iridescent black. Italian fishermen known as *spadularu* catch its close relative, the silver scabbardfish or 'spatola' *Lepidopus caudatus*. The fish can be seen in Sicilian and Calabrian fish markets around the Straits of Messina where there is deep water and upwellings that bring twilight zone creatures closer to the surface.

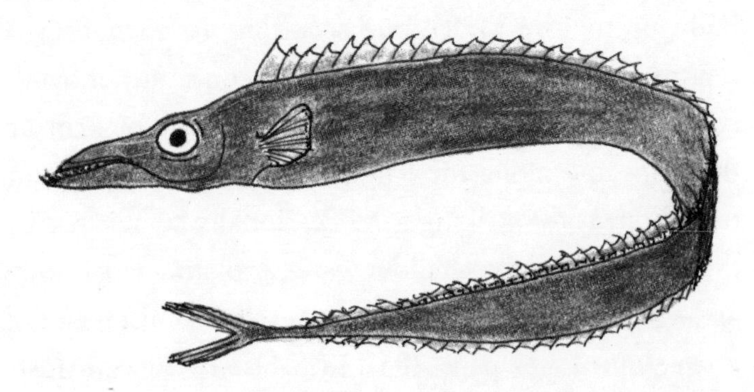

Black scabbardfish

The extraordinary barrel-eye fish *Macropinna microstoma* appears to have a transparent forehead. Its huge, tubular eyes capped with bright green lenses rotate inside the transparent, fluid-filled shield that covers the fish's head. When it is searching for prey in the ocean overhead, the eyes point upwards, and when it is feeding, the eyes rotate to face forwards. Two structures above the small, pointed mouth, which look like tiny eyes, are actually nostrils. The fish is thought to be partly a kleptoparasite in that it probably steals prey from other creatures, such as 10m (33ft) long, colonial, jellyfish-like animals called siphonophores that float about in the deep sea like living drift nets. The barrel-eye manoeuvres carefully among the siphonophore's tentacles, the transparent shield protecting its eyes, and steals copepods and other small crustaceans the creature has trapped.

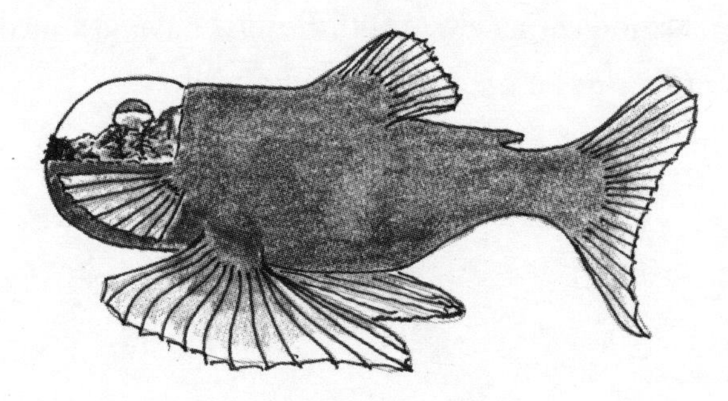

Barrel-eye fish

A small siphonophore *Erenna* – just 45cm (18in.) long – which lives deep in the Pacific is the first marine invertebrate known to use a bioluminescent lure. It glows red to catch its dinner. This type of siphonophore is divided into two sections: the upper half consists of a chain of pulsating bells, like miniature jellyfish, that pull the entire creature along like a freight train, while the lower half is the food-catching section, with batteries of tentacles that are not only armed with stinging cells, but also with tiny 'tentilla' that branch off from some of the tentacles and are tipped with red, glowing blobs. The shape of the red tips closely resembles that of deep-water copepods, one of the principal foods of mid-water fish, and they are twitched to mimic the copepod's erratic swimming pattern. It seems fish are attracted to the blobs but are zapped by the stinging cells and engulfed by the siphonophore. This was confirmed by observations by scientists from the Monterey Bay Aquarium Research Institute, who found both small fish and red blobs in some of the siphonophore's many stomachs, indicating that the fish must have grabbed the red tentilla tips before it was caught itself.

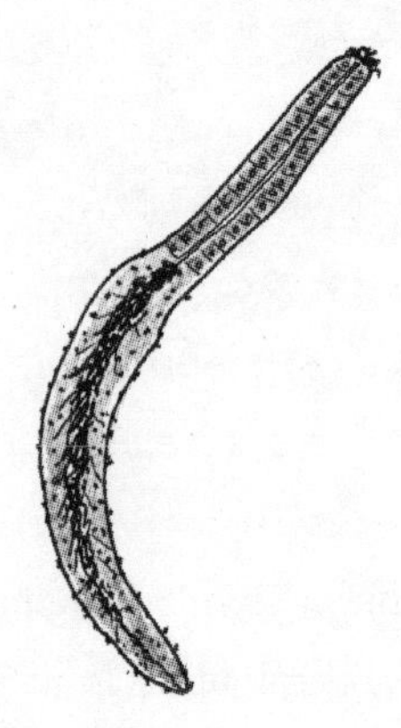

Siphonophore

IN THE INKY DARKNESS OF THE MIDNIGHT ZONE

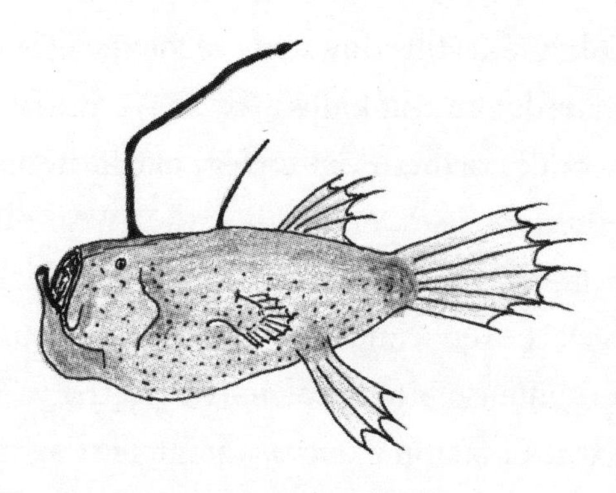

Krøyer's deep-sea anglerfish

The female of one type of fish does not leave finding a partner in the vast darkness of the deep sea to chance; she carries him with her. The small male Krøyer's deep-sea anglerfish or sea devil *Ceratias holboelli* embeds himself in the female's body and lives there as a parasite, ready to fertilise her eggs on demand; to ensure breeding success she might have half a dozen males attached to her. The female is the biggest of the deep-sea anglerfish, the largest individuals over a metre long, and she is a formidable predator with a huge mouth in front of which she waves a glowing lure on the tip of a fishing rod, a modified dorsal fin spine. The bioluminescence is produced not by the fish itself but by bacteria stored in the lure. When prey approaches to investigate she grabs it and swallows it whole; in fact, her jaws and stomach can stretch so much she can swallow prey twice her size. She hunts at depths of 1,000–4,400m (3,281–14,436ft).

About 1.6km (5,250ft) down in the Pacific Ocean, at the edge of an undersea cliff known as the Gorda Escarpment off the shore of northern California, are the nests of deep-sea fish and octopuses. A rare find, they were discovered by the cameras of a remotely operated vehicle of the Monterey Bay Aquarium Research Institute in a localised area of intense biological activity in the vast void of the deep sea; in fact, it's one of the highest aggregations of fish and octopuses ever found in the depths. The nesting fish are blob sculpin *Psychrolutes phrictus* and the nests of the octopus *Graneledone* are located within 5m (16ft) of the fish nests, both of which are close to cold seeps (see page 78) with tubeworms and clams. The sculpin nests are within 1–2m (3.3–6.6ft) of each other and each consists of a mass of 100,000 large pink eggs. They are free of dirt, indicating that the parents keep them clean and guard them, the first evidence of parental care in a deep-sea fish. The octopuses brood eggs under their bodies and sit motionless with their arms curled. It seems part of the reason for nesting on the edge of the escarpment is that the current is strong, bringing oxygen to the nest sites and wafting away wastes, and when the fish and octopus hatchlings are ready to emerge, it carries them away to be distributed elsewhere in the ocean.

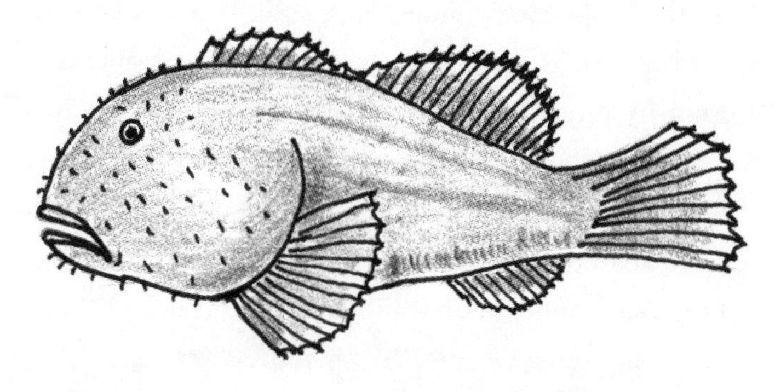

Blob sculpin

The deep-sea pelican eel *Eurypharynx pelecanoides* is a swimming mouth. Its enormous maw is far wider than its long and slender metre-long body. The lower jaw is like a pelican's pouch, hence the common name, and it is loosely hinged so the pelican eel can catch and eat fish considerably larger than itself. But it is something of an enigma. Eel stomachs tend to be filled with deep-sea crustaceans, small fish and squid rather than large fish, and it has a curious organ on the tip of its tail that has all the trappings of a lure, which glows pink with occasional flashes of red, but at the wrong end of its body! Its function must remain a mystery for now. However, according to Danish scientists, who studied 760 specimens caught in the Atlantic Ocean, the mouth does not suck in prey anyway, but is pushed forwards, whereupon it expands, much like the mouth and throat of a baleen whale. The eel has forward-pointing eyes, indicating stereoscopic vision, and a sensitive lateral line system that projects from the body rather than being sunk beneath the skin as with other fish – both valuable senses for obtaining a good fix on a target in the dark. Even so, the pelican eel seems to be somewhat random in what it catches: many specimens were found to have Sargasso weed in their stomachs.

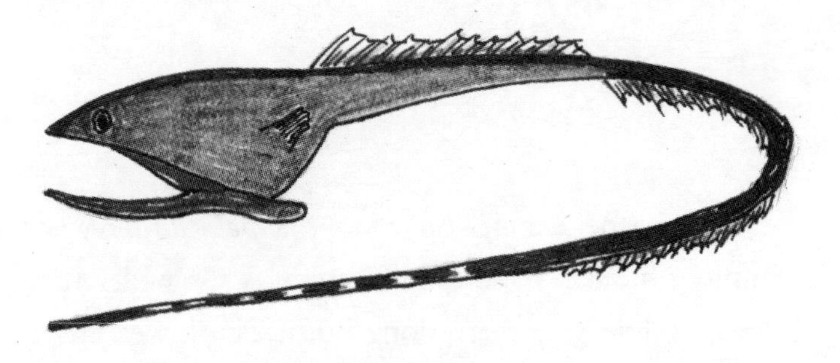

Pelican eel

The scaleless dragonfish *Grammatostomias flagellibarba* is only 15–20cm (6–8in.) long, and the whisker-like barbel under its chin, tipped with a light-producing organ that is waved about as a lure, can be ten times the length of the fish itself. Yet, despite its small size, it is a ferocious predator. Compared to its body size, it has exceptionally large fang-like teeth. It has rows of bioluminescent organs along its lower sides, which it can switch on and off at will. If it should catch something, the walls of its stomach are coloured black and its lights concealed, so it is almost invisible until it has digested its meal and is actively hunting again. It lives in the North Atlantic, including in the Bay of Biscay and waters off southern Ireland.

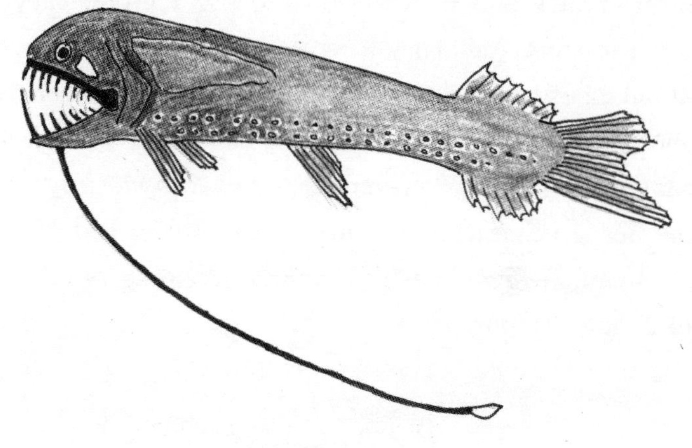

Scaleless dragonfish

Sea cucumbers in shallow waters resemble leathery sausages and have rows of tube feet on the underside (like their relatives the starfish), with which they are able to crawl extremely slowly on the seabed. Those living in the deep sea, however, can swim and are surprisingly agile. Some, such as *Enypniastes*, have webbed swimming structures at both ends. With these they can leave the sea floor and rise as much as 1,000m (3,281ft) in the water column. When researchers from Oceaneering International first saw the creature in 2007 at a depth of 2,500m (8,202ft), they nicknamed it 'the headless chicken fish', because seen from certain angles it bears a striking resemblance to a headless chicken.

Acorn worms feed on bottom sediments, such as those on the floor of the abyssal plain. They are not 'worms' as such, but the evolutionary 'missing links' between invertebrates and backboned animals, a group called the hemichordates. They have no eyes or obvious sense organs, but they do have a head end, with a proboscis and collar, and a tail end, and they can swim, albeit in a rudimentary way. Wherever deep-sea acorn worms feed, they leave a characteristic spiral trail on the seabed. Most are small, 9–45cm (3.5–18in.), but one species can grow to 2.5m (8ft) long.

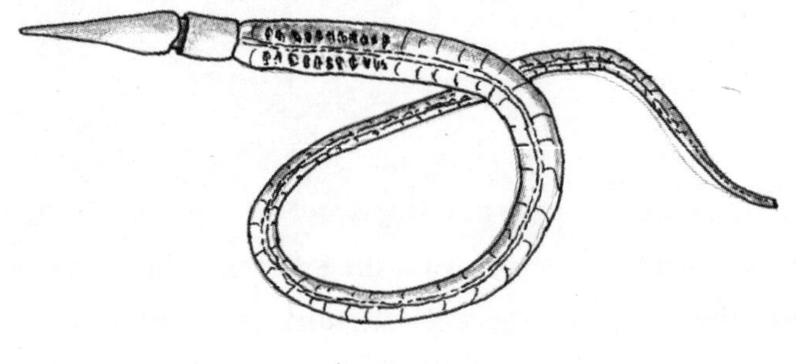

Acorn worm

The transparent amphipod *Phronima* is a hitch-hiker from hell. The female's entire head is covered by huge compound eyes that look left, right and upwards in search of movements in the near darkness. She is looking for salps. She catches them and then eats their insides to leave a hollow barrel that becomes a nest for her eggs and a nursery for the hatching larvae. She propels the salp husk through the water so her youngsters are bathed in fresh food and water.

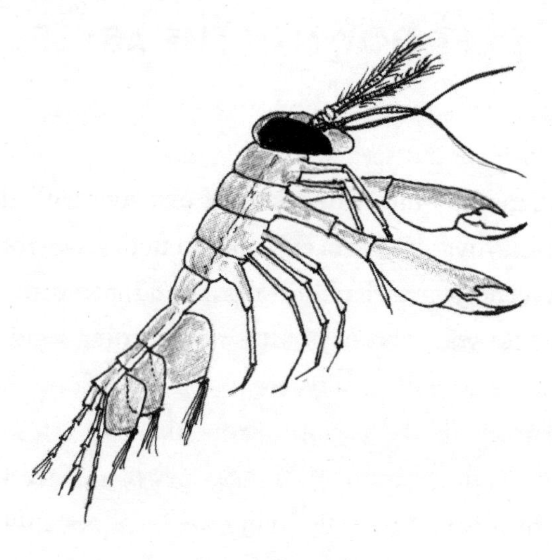

Deep-sea amphipod

DEEP DOWN IN THE ABYSS

This zone is one of the largest environments for life on Earth, yet it is the least populated. It covers about 60 per cent of the Earth's surface, 83 per cent of the oceans and seas, and originates in the polar regions as a continuous flow of cold water that moves along the deep seabed towards the equator from the poles. It's a very calm place, divorced from the rigours of surface waters, which means the animals living here are not influenced by changes in day and night or the seasons, so they can be delicate in structure.

The 30cm (12in.) long tripod fish *Bathypterois grallator* literally perches on the muddy deep-sea ooze. It has extra-long fin rays, up to a metre long, on the pectoral and tail fins that, as its name suggests, form a tripod. With these it sits down on the bottom, facing the current, waiting for small deep-sea marine life, such as copepods, to drift in. It's a curious fish in that it is a true hermaphrodite, for it has both male and female sex organs, and they both mature at the same time so if it doesn't find a partner it can fertilise itself – a useful trick to have up your sleeve in the vastness of the deep sea.

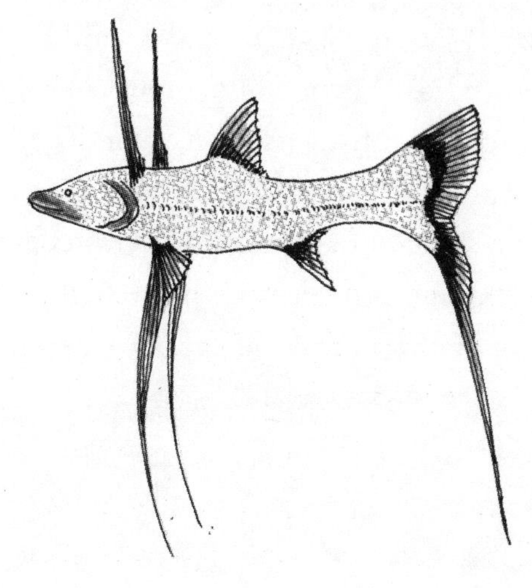

Tripod fish

The black swallower *Chiasmodon niger*, like many deep-sea fish, has an enormous stomach, so it can swallow fish much larger than itself – twice its length and ten times its mass. It's only 25cm (10in.) long at most, but has a mouthful of interlocking teeth that 'walk' from the prey's tail to its head until it is firmly coiled in the predator's stomach. This means that nothing can escape but it can be a handicap for the black swallower: it can take prey that is far too big for its digestive system to cope with. Decomposition sets in before the fish has had time to digest its meal and the release of decomposition gases forces the fish to the surface, which is the way most specimens have been collected. In fact, a black swallower found floating on the surface off the south coast of Grand Cayman in the Caribbean in 2007 had swallowed an 86cm (34in.) long snake mackerel, a very aggressive fish in its own right. The swallower, however, was just 20cm (7.9in.) long. The prey, over four times as big as the hunter, lay coiled in the black swallower's stomach.

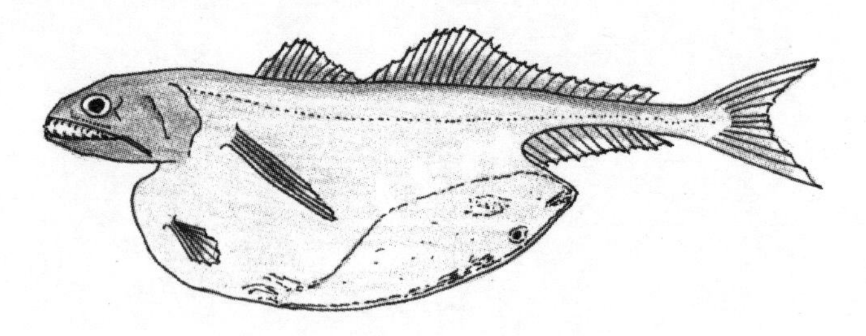

Black swallower

Tunicates are basically transparent bags that take in water at one end, filter out food particles and expel it at the other end. The predatory tunicate *Megalodicopia hians* is different. It looks like a transparent ice-cream scoop attached to the walls of submarine canyons or the deep-sea floor down to depths of 5,000m (16,404ft), its most distinguishing feature being what looks like a huge and cavernous mouth. The mouth is formed from the edges of the incoming siphon, which form 'lips'. It is sensitive to the vibrations of tiny shrimps and copepods in the water and the 'mouth' closes on them like a Venus flytrap. This predatory, albeit incredibly slow, lifestyle enables this tunicate to occupy oxygen-depleted places in the ocean that are less productive.

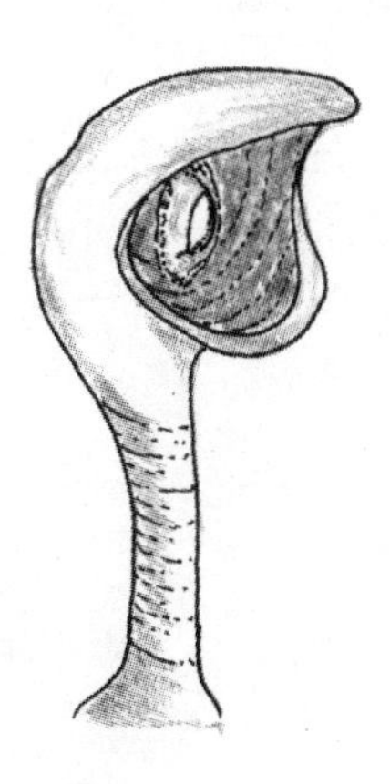

Predatory tunicate

The 'Dumbo octopus' *Grimpoteuthis* has a bulbous body and large fins that resemble the ears of Walt Disney's famous flying elephant. There are several species living at depths ranging from 400m (1,312ft) to 4,800m (15,748ft), with the suggestion that some can be found even deeper in the next zone down, which would be the greatest depth known for any octopus. They grow generally to about 20cm (8in.) long, but one specimen hauled up in 2009 from the North Atlantic measured 1.8m (about six feet) long. These octopuses have large eyes, occupying about one-third of the head, and a primitive, internal, U-shaped cartilage 'shell' that is thought to support the fins, a feature unknown in other octopus species. They hover over the sea floor and consume deep-sea crustaceans, worms and shellfish, which they swallow whole, unusual feeding behaviour for an octopus.

Dumbo octopus

IN THE DEEPEST DEEPS

When Hollywood film director James Cameron piloted his submersible to the bottom of the world's deepest trenches during the Deepsea Challenge expedition in March 2012, he did more than break records. Cameras on the craft revealed a surprising wealth of marine life at great depths. On the way down into the Challenger Deep in the Mariana Trench – the world's deepest at 11km (6.8mi) below the surface – he saw giant, single-celled, amoeba-like creatures known as xenophyophores (see below), along with giant shrimp-like amphipod crustaceans, marine relatives of woodlice or pillbugs that are 17cm (7in.) rather than the more usual 1–2cm (less than an inch) long. Stalked sea anemones on pillow lava dominate parts of the New Britain Trench, near Papua New Guinea, while shallower waters have spoon worms, small burrowing creatures that create a clear circle around their burrow entrances when using their proboscis to lick organic matter off the surrounding sediment.

Xenophyophores are extraordinary creatures, and unfortunately they do not have a more manageable common name. They are essentially huge single cells – great blobs of living cytoplasm in which are embedded many cell nuclei. One species, with the equally tongue-twisting name of *Syringammina fragilissima*, is 20cm (8in.) in diameter, making it one of the largest known single-celled organisms. Xenophyophores live deep down in the ocean. Many inhabit the abyssal plain where they root through the muddy sediments, and some, spotted in the Sirena Deep of the Mariana Trench, are present at depths down to at least 10,641m (34,911ft) – but they could be living even deeper. They feed like amoebas, enveloping food with a foot-like lobe of cytoplasm, known as a pseudopodium. While feeding they produce slime, and in the bottom of ocean trenches the slime together with waste matter can cover large areas. They appear to be very important in deep-sea ecology, and several other deep-sea creatures appear to associate with them, including isopod crustaceans, brittle stars and worms, but the relationships are little understood. Aside from this, little is known about xenophyophores for they are very delicate and break up easily when caught.

The fish discovered living at the greatest depth was a cusk eel *Abyssobrotula galatheae*, dredged up from a depth of 8,372m (27,467ft) in the Puerto Rico Trench in 1970. It was already dead by the time it reached the surface. However, in 2008 a UK–Japan team, using remote cameras, observed a shoal of fish at a depth of 7,703m (25,272ft) in the Japan Trench, the greatest depth at which living fish have been seen. They were seventeen 30cm (12in.) long snailfish *Pseudoliparis amblystomopsis*. It was always thought that creatures living at these depths would be relatively slow-moving to conserve energy, but the snailfish can be seen actively moving about in the darkness, feeding on deep-sea shrimp-like creatures which they detect with vibration receptors on their snouts.

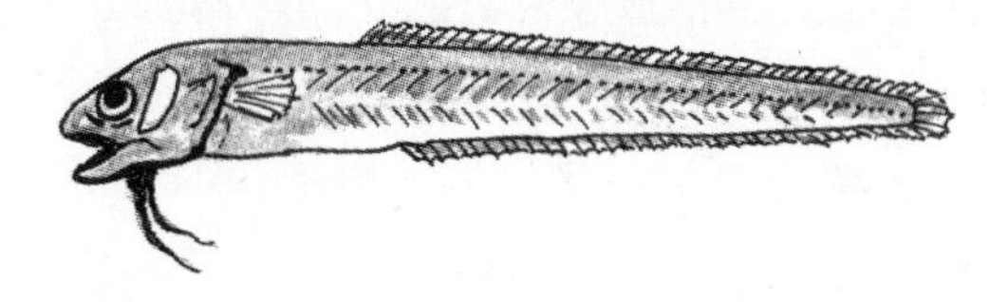

Cusk eel

It has long been thought that the very bottom of the deepest ocean trenches is devoid of life, but Scottish researchers sent a remote sampling device to the floor of the Mariana Trench and revealed that the sediment down there is filled with microbes. They are twice as active as those found at shallower sites and feed on dead organisms sinking down from the surface.

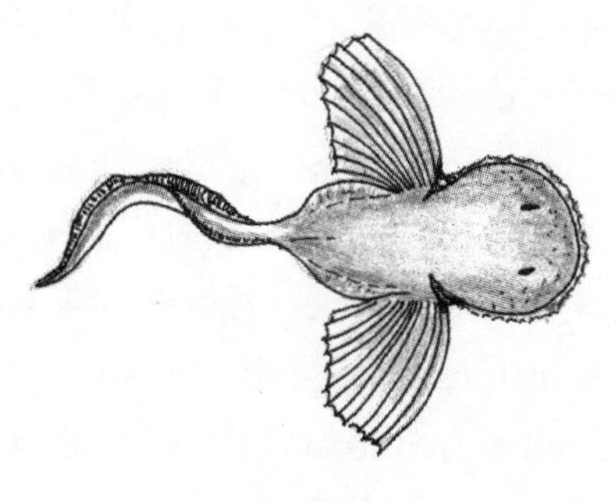

Snailfish

RAINING DOWN FROM ABOVE

Vampire squid

The vampire squid *Vampyroteuthis infernalis* is an enigmatic cephalopod living between 600m and 900m (1,969–2,953ft) down and probably deeper. It does not behave like any other squid, cuttlefish or octopus. For starters, this small, 30cm (12in.) long squid looks like a cross between a squid and an octopus, with its arms joined by webbing and lined on the underside with fleshy spines. Its entire body is covered with light organs, and it varies in colour between reddish and black velvet. Its beak is white and its enormous eyes are generally red, though sometimes appear blue depending on the lighting. Each eye is 2.5cm (1in.) across, which makes it the world's largest known eye for the size of the body. The squid moves with tiny fins, but its special feature is its ability to survive in very low oxygen conditions, in the so-called 'shadow zones' (see page 47) that can be found in this part of the ocean.

Adaptations include a low metabolic rate, blood proteins that transport oxygen more efficiently than those in other squid and octopuses, and a large gill area. Like several deep-sea squid, it lacks an ink sac; instead, when threatened it turns itself inside out, confronting the attacker with the rows of spines, and exudes a cloud of glowing slime containing many orbs of blue light from its arm tips. This dazzles the predator, while the vampire squid escapes into the darkness.

A second adaptation to low-oxygen and nutrient-poor environments is its method of feeding. It is the only squid not to hunt its prey, for it feeds on detritus, such as the remains of salps, faeces, dead plankton and moulted skins of crustaceans that float down from above. It has a pair of retractable filaments, in addition to its eight arms, which are cast like fishing lines. The filaments collect random bits of waste and are wiped on the arms where the food is mixed with blobs of mucus secreted by the suckers, which the squid duly consumes.

About 2,865m (9,400ft) below the surface of California's Monterey Bay live two newly recorded species of marine worms in the genus *Osedax*, also known as 'zombie worms'. They have no eyes, mouth or stomach, and the biggest of them is no longer than 6.35cm (2.5in.) – not greatly inspiring until you discover that they feed on an unexpected food: the bones of dead whales. The sheer number of worms feeding creates a red carpet atop the carcass, the redness due to the red feathery gills they have at one end of their body. At the other they have root-like appendages that burrow into the whalebone, using a powerful acid to break it down. It is thought that bacteria living inside the worm digest the fats and oils in the bone, providing their host with the recycled nutrients. Even more intriguing is the fact that all the worms found by the researchers from the Monterey Bay Aquarium Research Institute were females, each containing masses of eggs. Closer inspection, however, revealed that the even tinier males live inside the females, with up to a hundred males inside a single female. The researchers liken these worms to weeds, such as dandelions, for they produce huge numbers of fertilised eggs which are wafted away by the ocean currents, their developing larvae ready to colonise another whale carcass some distance away on the deep-sea floor.

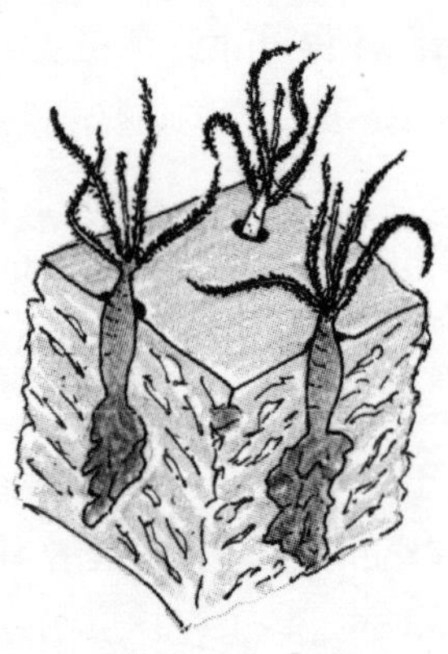

Zombie worms

Apart from mangroves and the odd palm, trees don't tend to grow in the sea, and certainly not in the deep sea, yet researchers from the Max Planck Institute for Marine Microbiology have found that trees washed into the sea, which have become waterlogged and have sunk to the deep-sea floor, became biodiversity hotspots, including acting as staging posts for animals that live at hydrothermal vents and cold seeps (see below). Wood-boring molluscs *Xylophaga*, known as 'shipworms', start to break down the wood and bacteria continue the process. The bacteria produce sulphides and these are utilised by symbiotic bacteria that live inside vent animals: they manufacture food for their hosts. Where vents are widely spaced on the seabed, sometimes by hundreds of kilometres, the wood provides a 'half-way house'. One surprising animal also to be found eating wood is a squat lobster *Munidopsis andamanica*. Researchers from the University of Liège in Belgium discovered that the lobster has bacteria and fungi in its gut that break down wood, as well as other terrestrial plant material, such as leaves and coconut fragments, which have sunk to the deep-sea floor.

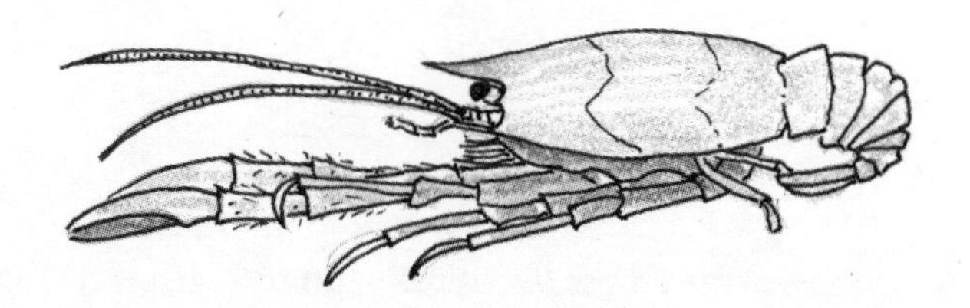

Tree-eating squat lobster

DEEP-SEA HOT SPRINGS

Hot springs on the deep-sea floor are known as hydrothermal vents, and life here in the dark depends not on energy from the sun, like most of the rest of life on Earth, but from heat at the centre of our planet. Bacteria living in the warm water feed on hydrogen sulphide and are at the bottom of a food web that includes giant clams, ghostly white crabs, squat lobsters, eelpout fish, eyeless shrimps and bright red tubeworms.

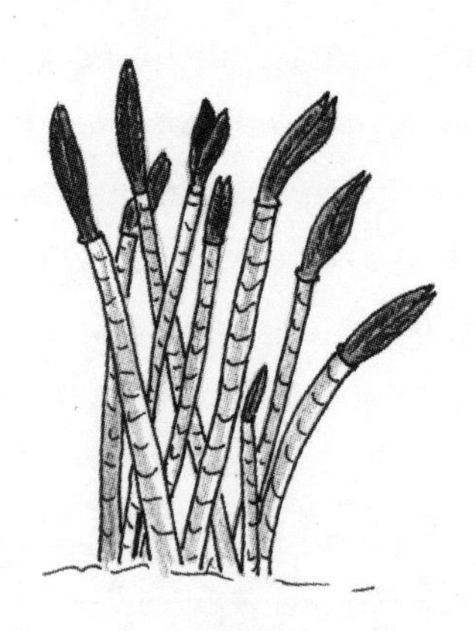

Giant tubeworms

Giant tubeworms *Riftia pachyptila*, up to 2.4m (8ft) long, are the most spectacular hydrothermal vent animals.

They live on the floor of the deep sea close to black smokers – chimney-like structures that spew out hot water at 400°C (752°F) containing black mineral particles, especially sulphides. The tubeworms have a bright red plume at the top, which acts like gills, and this can be retracted into the tube if danger threatens. Symbiotic bacteria living in a special pouch in the worm's body use hydrogen sulphide, carbon dioxide, oxygen and other chemicals absorbed by the worm to manufacture food for their host, and they must be efficient: the worm is the fastest-growing invertebrate known. It can reach a length of 1.5m (4.9ft) in just two years.

Scaly-foot gastropod

The scaly-foot gastropod *Crysomallon squamiferum* is a deep-sea snail that lives close to the base of black smokers. It is unusual in having a foot that is armoured with scales containing iron minerals. Similarly, the outer layer of its shell is made of iron sulphides, while the middle layer is organic and the inner layer made of aragonite, a form of calcium carbonate found in the shells of other molluscs. This three-layered shell is especially tough, the organic layer being able to absorb the strain of the squeezing action of, say, a crab's claws. Moreover, it is the only animal on the planet known to utilise iron sulphides in this way, and the US military are studying it with the development of new types of armour in mind.

The 13cm (5in.) long Pompeii worm *Alvinella pompejana* lives very close to hydrothermal vents. Its tail rests in scalding-hot water, up to 80°C (176°F), while its head is in cooler water at 22°C (72°F). Its head has red feathery gills, and its body is covered with a hairy grey 'fleece' of bacteria that helps insulate its body from the heat. Aside from tardigrades or 'water bears', which can survive temperatures of 150°C (302°F), the Pompeii worm is the most heat-tolerant complex animal on Earth.

Pompeii worm

The yeti crab *Kiwa hirsuta* is covered in bacteria too. It lives on hydrothermal vents 2,200m (7,218ft) down in the Pacific Ocean 1,500km (932mi) to the south of Easter Island. Its pincer arms are covered with what looks like fur, and living among the hairs are filamentous bacteria. Their function is unknown, but the crab itself (actually a type of squat lobster distantly related to hermit crabs) is so unusual it has been classified in the family Kiwaidae, an entirely new family all to itself.

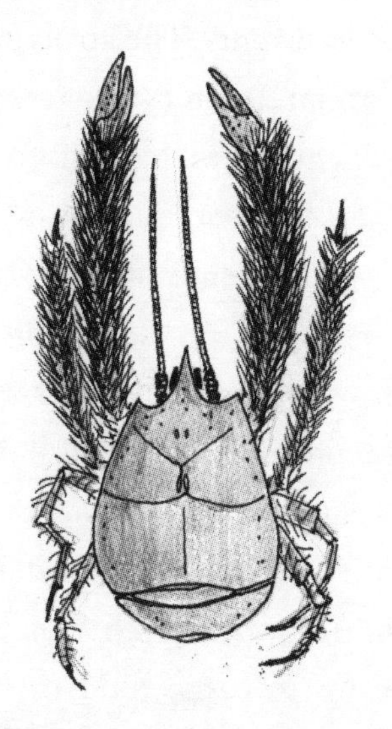

Yeti crab

DEEP-SEA ANOMALY

On mid-ocean ridge systems in both the Atlantic and Pacific oceans scientists have found a very distinctive but peculiar pattern in sediments on the sea floor close to hydrothermal vents. The structure is shaped like a convex shield, about 10cm (4in.) across, with rows of tiny holes and vertical shafts that link to a matrix of tiny tunnels below the sediment. The tunnels form a regular pattern, each at an angle of 120° to its neighbour. Despite having scooped up sediment containing the burrows, scientists have yet to discover what causes them or even what lives in them; yet they have already given the mysterious creature a name: *Paleodictyon nodosum*. There are several suggestions, including: 1) the pattern is caused by a marine worm with a burrow system that either deflects water to catch food or builds a kind of farm in which the animal rears its own food (like a leaf-cutter ant growing a crop of fungus); 2) it is created by a giant single-celled xenophyophore (see page 65); or 3) it is made by a type of sponge that has taken to living in the sediment. Whatever the modern inhabitant, it was not the first to adopt such a system, for Eocene fossils, about 50 million years old, show a similar pattern. They were first discovered in Spain in the 1950s and later in rocks throughout Europe, some as old as 500 million years, which would make this honeycomb-like pattern one of the earliest known deliberate structures made by a living thing. Interestingly, the younger the rocks the more mathematically precise are the patterns.

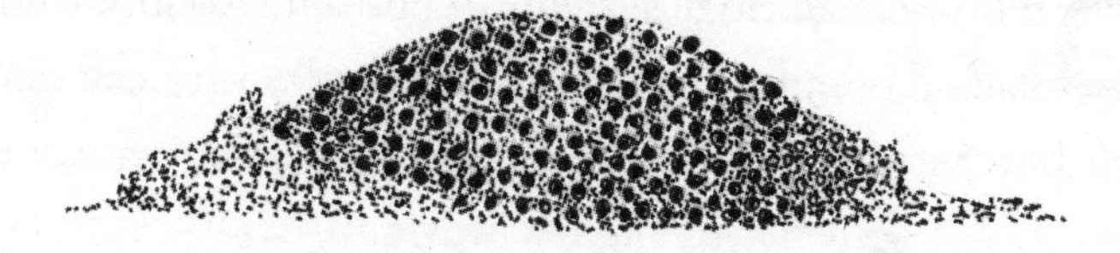

Sediment pattern of unknown animal

COLD SEEPS

Cold deep-sea hydrocarbon seeps are cracks in the seabed, similar to hydrothermal vents except that the mineral-rich fluids escaping are at the same temperature as the seawater or slightly warmer.

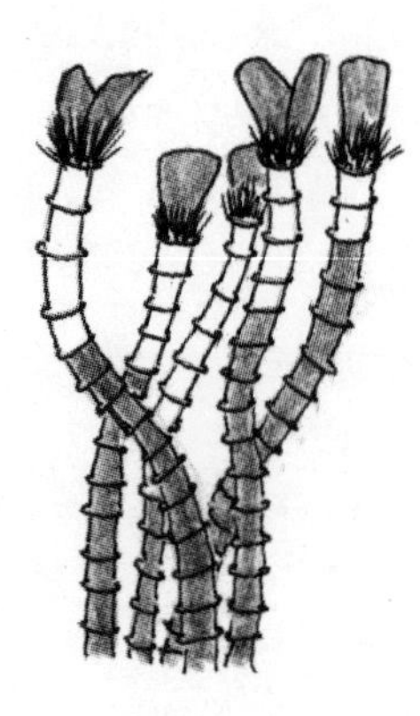

Beardworms

Bacteria mats grow around cold seeps, along with beds of mussels and clams that do not feed like other shellfish but rely for energy on bacteria living in their tissues that use methane and hydrogen sulphide from the seep. The microbial activity results in deposits of calcium carbonate, which become home to beardworms (Family:

Siboglinidae) and other types of giant tubeworms, which not only extract sulphides from the water, but also have a kind of 'root' system that bores into the sediment to extract the chemicals.

One species of tubeworm *Lamellibrachia luymesi* lives in bush-like aggregations close to cold seeps on the seabed at a depth of 800m (2,625ft) and grows extremely slowly. Unlike its fast-growing hot vent relatives (see page 72), these worms can take more than 250 years to reach a greater length of 3m (10ft), making them the longest tubeworms and among the oldest known non-colonial animals without backbones (see page 21). They rely on symbiotic bacteria in their tissues to supply them with food.

Cold-seep tubeworms are either male or female, and in some locations a large parasitic bivalve mollusc *Acesta bullisi*, about 11cm (4.3in.) long, has latched onto the tube openings of female *Lamellibrachia* tubeworms, where it feeds on the fat-rich eggs as they are released. The mollusc's unusually shaped shell wraps around the worm's opening in such a way that the eggs are delivered straight into the bivalve's inhalant siphon, so few escape.

Some seep sites form 'brine pools' in depressions in the seabed, which have a salt content that is up to five times higher than the surrounding seawater. They look like deep-sea lakes, some up to 20km (12mi) across, for the brine does not mix with the overlying water, creating a distinct surface, shoreline and underwater waves if a submersible should touch them. High concentrations of methane provide the chemical energy for bacteria, and beds of deep-sea seep mussels, with bacteria in their tissues, grow on the edge of the pool, like mussels on the seashore.

BELOW THE SEABED

While most marine biologists are studying animals living in the water or sitting on top of it, there are a few dedicated researchers who are looking at the rocks and sediments that make up the deep-sea floor itself for signs of life. Extraordinary as it may seem, they are finding it, and lots of it.

In red clay sediments on the deep ocean floor of the Pacific Ocean, some layers many kilometres thick and possibly 86 million years old, live microbes that grow so slowly and use such minuscule quantities of oxygen that they can be barely considered alive. Cell division may take hundreds and, in some cases, thousands of years to complete, and these organisms will not have been exposed to food from the outside world since the dinosaurs ruled the planet. They tick over on the barest minimum of energy requirements for a living organism to exist. What's more, it is thought that 90 per cent of the Earth's single-celled organisms lie buried here in these deep-sea sediments.

Another piece of research focused on the rocks below the ocean-floor sediments at a site to the west of Washington state. Here, 3.5-million-year-old basalt rock is heavily fractured so seawater flows through the labyrinth of cracks. Several previous research projects have hinted that microbes probably live in those cracks; but bringing up samples from below the deep-sea floor is fraught with problems, especially contamination from organisms picked up from elsewhere in the sea. The international team confronted this challenge head-on and brought up samples in sterile conditions. The team grew the microbes at their natural temperature of 65°C (149°F) in a laboratory environment designed to 'resemble the chemistry of water flowing through the oceanic crust' for several years. They found that the energy keeping these microbes alive comes from the reaction of water in the cracks with chemicals in the rocks themselves. This produces the raw materials that, in turn, these methane- and sulphur-cycling bacteria use to manufacture organic matter. It adds further weight to the notion that life growing in cracks in rocks under the floor of the ocean is supported by energy released from geochemical reactions, which sets apart this vast ecosystem (probably the largest habitable zone by volume on Earth) from most of the rest of life on Earth.

When studying sediments dredged up from the bottom of L'Atalante Basin in the Mediterranean Sea, researchers from Marche Polytechnic University and the Natural History Museum of Denmark had a big surprise. The area, which is at a depth of about 3,000m (9,843ft), is known as a 'deep anoxic hypersaline basin', which, put simply, means there is too much salt and no oxygen. Certain types of bacteria and other organisms can live and thrive in such conditions but it was thought that multicellular animals could not ... that is, until now. The researchers discovered three brand new species of Loricifera (tiny multicellular animals that were unknown until 1983 because they lie hidden by attaching themselves very firmly to gravel). They were living in the basin sediments, an oxygen-free environment in salt-saturated brine so dense it does not mix with the upper layers of water. The exciting thing about the discovery is that researchers can use these 1mm (0.04in.) long creatures to study what life was like on the early Earth, between one and two billion years ago, when anoxic conditions might have persisted in the deep sea and multicellular animal life was surprisingly slow to evolve.

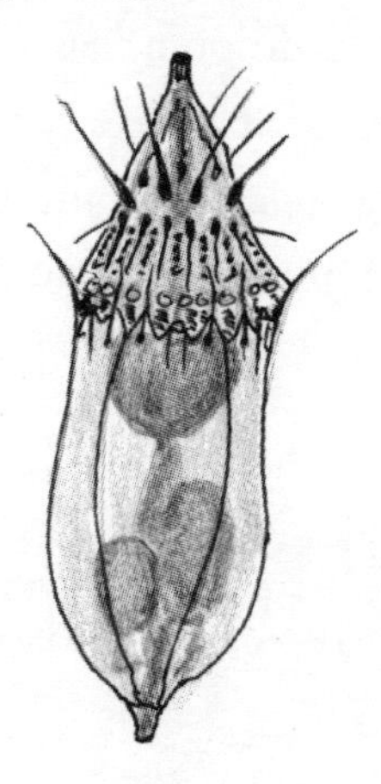

Loricifera

DEEP-SEA GIANTS

About 2,000m (6,562ft) down in the Pacific Ocean is an unexpected claim to the title of largest predatory shark. A true monster of the deep, the Pacific sleeper shark *Somniosus pacificus* grows, it is said, to 7m (23ft) long or more. It's both a hunter and scavenger, gliding with very little hydrodynamic noise to feast closer to the surface at night on giant octopus and squid, Pacific salmon, halibut and marine mammals, such as harbour seals *Phoca vitulina* and Dall's porpoise *Phocoenoides dalli*. Off the coast of Chile a female sleeper was caught and when her stomach was cut open, out fell a whole southern right whale dolphin! Similarly, the bluntnose sixgill or cow shark *Hexanchus griseus*, another deep-sea denizen and a relatively primitive shark whose close relatives are extinct, grows to lengths in excess of 5.5m (18ft).

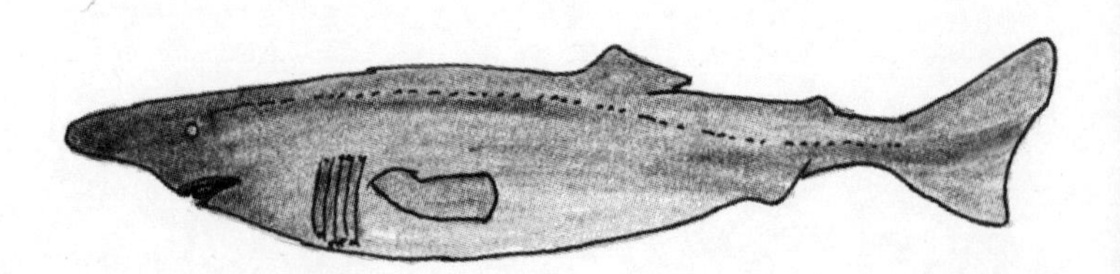

Pacific sleeper shark

Another giant among deep-sea sharks is the 6.4m (21ft) long and exceptionally slow-moving Greenland shark *Somniosus microcephalus*, which cruises sub-Arctic waters around Greenland and Iceland in the North Atlantic. It is a unique species in that it carries an unusual passenger. Attached to the cornea of one of its eyes is a parasitic copepod crustacean *Ommatokoita elongata* that glows in the dark. Speculation abounds that the bioluminescence attracts prey towards the shark's mouth, but this has yet to be proven. The shark itself is poisonous to eat, for the flesh contains intoxicating chemicals. For this reason, Inuit folk call a person who has drunk too much alcohol 'shark-sick'. However, it can be eaten if the flesh is boiled and the water changed several times, or if fermented by burying it in the ground, exposing it to repeated cycles of freezing and thawing over several months and then drying it. The fermenting process produces a fish dish that smells strongly of ammonia and is known as *kaestar hákarl*. It is said to be an acquired taste and popular with Icelandic people. It can also be made with basking shark.

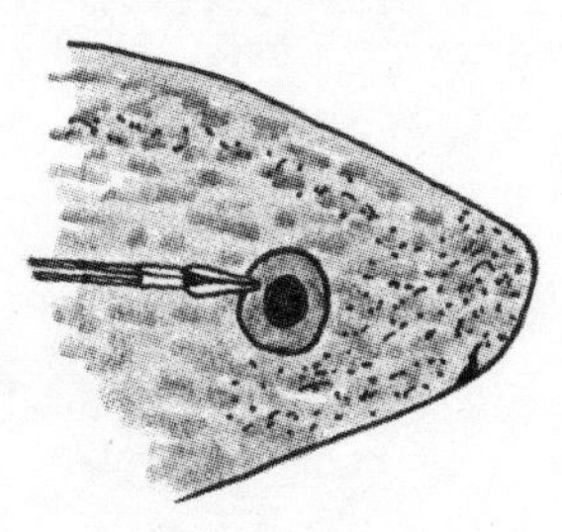

Greenland shark with a parasitic copepod

The megamouth shark *Megachasma pelagios* is another giant of the deep. It also grows to about 5.5m (18ft) long, but unlike its predatory and scavenging relatives it is a filter feeder, like the whale shark and basking shark (see pages 92 and 95). Its most distinctive features are thick, blubbery lips and floppy fins, indicating a big feeder and a slow swimmer. What is extraordinary about this fish is that it was totally unknown until 1976, when a specimen was trapped in a ship's sea anchor off Hawaii and hauled to the surface. How could such a giant go for so long without being seen by fishermen or recorded by scientists? Since then, only fifty-five or so specimens have been caught by fishermen or found washed up on beaches around the world and so the species has been little studied. A specimen caught off Los Angeles and released with a satellite tag gave a hint at how megamouth behaves. Like many deep-sea organisms it follows the daily vertical migration of zooplankton from the deep sea during the day towards the surface at night and back again. It feeds on deep-sea, shrimp-like crustaceans and jellyfish. Its discovery was probably the most significant in fish science in the twentieth century.

Megamouth shark

ODD SHARKS

The velvet belly lanternshark *Etmopterus spinax*, like the shark of the title, is another cutie; but unlike the epaulette shark, it lives in the depths of the ocean between 200m (656ft) and 1,000m (3,281ft), the bottom of its range being in the twilight zone where little sunlight penetrates. The shark is only 60cm (24in.) long at most, so it has to be careful not to fall prey to larger deep-sea predators. It does this in two ways, both of which involve the shark producing its own light. Firstly, it has light-producing organs or photophores along its belly. These help to camouflage it. Any predator passing below and looking up would see the shadow of the shark against the faint glow of sunlight at the surface, but with its photophores lit up the shadow disappears and the shark blends in with the background glow. This is a well-known phenomenon adopted by many small species of deep-sea sharks and other twilight zone fish. However, a second strategy has come as a surprise. Researchers at the Catholic University of Louvain in Belgium found that this little shark has spines on its back that are illuminated like *Star Wars*-style 'lightsabers'. In front of each of its two dorsal fins is a large spine, and photophores grouped around the base of the spines light them up or shine through them so that any predator approaching from above or from the side will see them from several metres away. The glowing spines in the inky darkness of the deep sea are saying, 'Don't eat me, I have spines on my back – I'm dangerous.'

Velvet belly lanternshark

The cookie-cutter shark *Isistius brasiliensis* may be small – no more than 50cm (20in.) long – but, size for size, it has the largest teeth of any known shark, and it uses them in an unusual way. It spends the day down in the darkness of the deep sea, but come the evening it heads for the surface, where it latches onto a dolphin, tuna or other large marine animal with its suction cup-like mouth and with its huge teeth takes a circular bite out of its victim's body. It was this that earned it the name 'cookie-cutter'. Like the lanternshark, it has light-producing organs along its belly, but it also has a band around its throat without illumination. It is thought this dark band looks like a small fish, and that the cookie-cutter uses it to lure potential victims closer, whereupon it shoots around and takes a cookie-sized bite from their flesh before disappearing into the dark.

Cookie-cutter shark

In 2001 a captive bonnethead shark *Sphyrna tiburo* (hammerhead species with the smallest 'hammer') gave birth to a single female offspring ... not unusual you might think, but the mother had not been in contact with a male shark for over three years and certainly not since she became sexually mature. It was the first recorded case of parthenogenesis or 'virgin birth' in sharks. In 2008, a blacktip shark *Carcharhinus limbatus* was found to have a viable embryo inside her even though she had been captured from the wild as a juvenile and isolated from males for over nine years. Then, in 2010, an international team kept eggs produced by a captive white-spotted bamboo shark *Chiloscyllium plagiosum* out of which hatched young bamboo sharks. The mother, however, had been reared from egg case to adulthood in captivity without any contact with males at all. These examples suggest that virgin birth is a possible shark survival strategy, with isolated groups of female sharks able to survive without males. It might be one reason why sharks have survived for hundreds of millions of years.

Bonnethead shark

Young scalloped hammerhead sharks *Sphyrna lewini* can get a suntan. The observation was made accidentally at the University of Hawaii when the pups were taken from the murky waters of Kaneohe Bay on Oahu (where an estimated 10,000 baby hammerheads hide from predators – especially bigger hammerheads – during the day), and placed in shallow ponds for study. The researchers found that after a few weeks their skin became much darker and, what's more, they tend not to suffer from skin cancers.

Scalloped hammerhead shark

The common thresher shark *Alopias vulpinus* is a fair-sized species, but it is mostly tail and so is not especially bulky. It can be up to 6m (20ft) long, but half of this is a very long, scythe-shaped tail, which it uses like a whip to stun its prey. By slapping the surface and even leaping clear of the water, it herds shoals of small fish and, in an extraordinary move for a shark, it swats them by thrashing its tail over the top of its head. It then quickly scoops up the dead and the dying. The pelagic thresher *A. pelagicus*, an open-ocean species, is known to come into shallower water to visit cleaning stations, such as that at Monad Shoal to the east of Malapascua Island in the Philippines. Here the sharks pass slowly or circle and bluestreak cleaner wrasse *Labroides dimidiatus* and moon wrasse *Thalassoma lunare* dart out to divest them of their parasites and dead skin.

Thresher shark

MONSTER SHARK

The biggest fish in the sea and the world's largest shark is the whale shark *Rhincodon typus*. The largest reliably measured specimen is generally accepted to be a shark caught off Baba Island, one of three tiny islands in Karachi Harbour, Pakistan, in November 1949. The shark was 12.65m (41.5ft) long, an exceptional size for a whale shark, but there have been claims of even bigger ones.

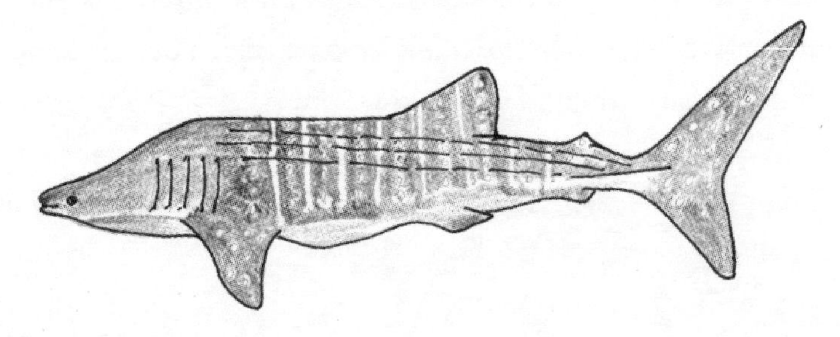

Whale shark

In the Gulf of Thailand a whale shark caught in a bamboo fish trap was estimated to be 17m (56ft) long and in Taiwan in March 1987 a landed whale shark was claimed to be 20m (66ft) long and 34 tonnes in weight. Off Honduras, where all whale sharks are known as 'Sapodilla Tom', there was an individual thought to be 21m (69ft) long;

and off the Yucatan peninsula another by the name of 'Big Ben' was close to 23m (76ft) long, but alas, no one ever had a chance to measure them accurately.

Such a huge animal, with muscles under its tough skin that can even stop bullets and repulse harpoons, is unlikely to have many enemies, but a chance find on the coast of the Gulf of Aqaba at the northern tip of the Red Sea hints at a contender. The frightening news was included in a letter from journalist and pioneer scuba diver Kendall McDonald to Gerald Wood, author of *The Guinness Book of Animal Facts and Feats*. It mentions the severed tail of a very large whale shark that was washed up on the shore of the Gulf in the early 1970s. Witnesses thought at first that the propeller of a large ship had cut it, but closer inspection revealed huge bite marks. What predator would have such a powerful bite is unknown, but it must have been pretty big!

Until recently, the life of whale sharks has been mostly unknown. They appear occasionally at the surface and then disappear into the depths. Where they go has been something of a mystery. Now, however, scientists are beginning to tease out the daily activity patterns of these giants. At night they hang out at the surface, but during the day they dive down to 1,000m (3,281ft) or more, a thick layer of fat below their skin providing insulation in deep waters that are not far above freezing. It is thought that, like many other marine animals, they are following the daily vertical migration of the zooplankton and other small organisms on which they feed. At certain times of the year, however, they break their normal swim pattern and head for places where there is a sudden and reliable food bonanza. Off the coast of Belize, for example, at the time of the full moon, they gather at traditional spawning sites along the country's extensive barrier reef (the second largest in the world after Australia's Great Barrier Reef), where shoals of fish, such as Cubera snappers *Lutjanus cyanopterus*, are shedding their eggs and sperm into the water. Here the sharks remain in shallower surface waters both day and night, vacuuming up the clouds of snapper caviar.

Giants in fairy tales are not known for their intelligence, but this ocean giant is turning out to be surprisingly bright. Whale sharks in Cenderawasih Bay, on the north coast of New Guinea, have learned how to latch on to fishing nets with their enormous letterbox-shaped mouths and suck out the shoals of small silverside fish, known locally as *ikan puri*, caught inside, thus stealing the catch of local fishermen. The fishermen put to sea in *bagans*, which are equipped with nets, platforms and underwater lights, but they do not discourage the sharks' piratical behaviour. Whale sharks bring good luck so they are encouraged, so much so that the sharks have become residents and the bay has been designated a Marine National Park.

BASKERS

The world's second largest fish feeds on the ocean's smallest creatures. Like the whale shark and megamouth, the basking shark *Cetorhinus maximus* is a filter feeder, gulping in plankton and other small organisms that accumulate in great numbers along the fronts between currents. It can grow up to 12m (39ft) long, the largest known specimen being one caught in a herring net in the Bay of Fundy in 1851. It was 12.2m (40ft) long, with a head 1.5m (5ft) across and a mouth about a metre (3.3ft) wide. More usually people encounter 'baskers', as they are known, that are 6–8m (20–26ft) long, the males being smaller than females. They are instantly recognised by their high dorsal fin, bulbous snout, cavernous mouth and huge gill slits. They turn up with unfailing regularity off the south coast of Cornwall each spring, and recent research has shown that they travel enormous distances across the ocean.

Basking shark

In the western part of the North Atlantic basking sharks quit the cooler waters off Cape Cod in the autumn and head south. Some end up in deeper waters off the Bahamas, in the Caribbean Sea and even as far south as Brazil, where a shark was tracked to the mouth of the Amazon. Baskers seen around the British Isles in summer also head for deeper waters in winter, but one individual made a most extraordinary journey. It was tagged in 2007 off the Isle of Man, and was found to have swum 9,589km (5,958mi) to Newfoundland, Canada.

The basking shark has come to be centre stage in sea monster mysteries. In 1977, for example, the Japanese fishing boat *Zuiyo-maru* hauled up the carcass of an unidentified sea creature from waters off New Zealand. The 10m (33ft) long animal looked like a plesiosaur, with a small head, long neck and huge body, but reputable scientists were quick to point out that a decomposing basking shark would fit the description. Its huge gills, which almost circle the throat, break down, leaving a small cranium (brain box) on what appears to be a long neck. Similarly, the famous 'Stronsay Beast', which was washed up after a storm in the Orkney Islands in September 1808, was a basking shark that had decomposed in the same way.

WHITE DEATH

Since the phenomenon that was *Jaws* hit the world, mention of the great white shark *Carcharodon carcharias* conjures up spine-chilling images of razor-sharp teeth set in the protrusible jaws that chomp on just about anything with warm blood, including us. It is known by a variety of common names – white pointer, blue pointer, white shark, uptail, Tommy shark, man-eater and white death – and wherever it is encountered it is feared. It is a stealth predator, sneaking up from below and behind its victims. If it can, the shark will swim over dark rocks rather than a pale sandy seabed so it is less likely to be seen from above. When it gets close enough it rushes towards the target to take the first exploratory bite. However, the shark is travelling so fast that if the prey jigs to one side, it cannot stop and shoots straight out of the water. On the other hand, if it makes contact and likes what it has found it will continue to feed, its bottom teeth holding the prey like a fork and the top teeth slicing through it like a knife.

Great white sharks are considered to be the largest predatory sharks in the ocean. Large adult sharks average about 4.6m (15ft) these days, but they can grow considerably larger. During the eighteenth, nineteenth and early twentieth centuries, sharks caught in the Mediterranean (see page 167) and off South Africa and New South Wales were said to have exceeded 6m (20ft), one of the Australian sharks having razor-sharp, triangular serrated teeth 6.4cm (2.5in.) long. Two sharks found in New Brunswick on Canada's Atlantic coast, however, were even bigger. The largest was thought to be 11.3m (37ft) long. It was found trapped in a herring weir at White Head Island, as was a 7.9m (26ft) specimen found under similar circumstances at Harbour de Loutre on Campobello Island. Not far away in the Gulf of St Lawrence a 7.6m (25ft) long great white was stranded. Canada, it seems, has been host to some pretty big sharks. Not to be outdone, California had its own monster. Local fishermen used the length of their boats to judge its size and estimated it to be in excess of 9.5m (31ft) long. Needless to say, no one dared to harpoon it.

Great white shark

Great white sharks appear to change their food preferences with age. While they will take any food that happens to be in front of them, generally juveniles catch fish, while older sharks switch more to sea mammals, especially seals and sea lions. The largest mature individuals – 4.5m (15ft) long or more – have a penchant for whale blubber. Watching great whites off South Africa, Apex Expeditions and the University of Miami collaborated to study sharks attracted to whale carcasses. As soon as a carcass appeared, they found that the larger sharks – sometimes as many as forty at the same time – abandoned their local seal hunt and gathered around the floating, bloated whale. This did not trigger a feeding frenzy, however, for great whites seem to have immaculate table manners, and it's all down to size.

The largest sharks feed first, sometimes alongside each other without signs of aggression, while the smaller individuals wait patiently in the queue and are usually confined to the less desirable cuts, often pieces of blubber floating in the water. The larger sharks eat the whale's tail flukes first, before chomping on blubber elsewhere on the body. One 4m (13ft) long shark was seen to pull a near-term foetus from the uterus of a dead Bryde's whale *Balaenoptera edeni* and consume it. Sharks, it seems, can get enough calories from a 30kg (66lb) chunk of whale blubber to keep going for at least six weeks. They will even regurgitate low-energy meat and go back to feed on chunks of high-energy blubber. When sated, the large sharks seem to go into a postprandial torpor, touching the whale with their snout and open mouth but not biting, as if the best pieces have gone and it is not worth eating any of the rest.

Interestingly, the researchers noticed that the largest sharks at nearby Seal Island, a focus for predatory attacks on Cape fur seals *Arctocephalus pusillus*, were no bigger than 3.5m (11.5ft) long. Larger sharks were seen less frequently; in fact, they constitute less than 14 per cent of the observed population. It is as if they have grown too big and cumbersome to outwit and out-manoeuvre the agile seals, and that a mouthful of seal was insufficient to satisfy a large and hungry shark anyway. However, as soon as a whale carcass appeared, so did the 4.5m (15ft) sharks, with some giants over 5m (16ft) long. Southern right whales *Eubalaena australis* that over-winter around False Bay in South Africa and an inshore population of Bryde's whales are more likely to be hit by ships or caught in fishing gear than whales out at sea, providing ample food for the ocean's largest predatory fish.

Great white sharks were once thought of as stay-at-home creatures but all that changed when a 3.8m (12.5ft) long mature female left her South Africa home at Dyer Island and headed out into the Indian Ocean. She was nick-named Nicole (after Nicole Kidman) by the researchers from the White Shark Trust and local universities and institutions, and on 7 November 2003 they attached a pop-up satellite tag to her. On 28 February 2004 the tag released itself from the shark, by which time Nicole was 2km (1.2mi) from the shore and about 37km (23mi) south of Exmouth in Western Australia. The journey across the entire Indian Ocean had taken just ninety-nine days. Even without her tag, Nicole was a shark easily recognised by her distinctive dorsal fin, so imagine the surprise when she was spotted again back in South Africa on 20 August 2004. She had taken nine and a half months to complete the 20,000km (12,427mi) round trip, which meant a minimum speed of 4.7km/h (2.9mph) – one of the fastest transoceanic migrations of any known marine animal (see page 38). Why she makes the journey is unknown.

Great white sharks have been found to make unexpectedly long journeys in the eastern Pacific too. Sharks tagged off California and Mexico have been found to head out into the open ocean and mill about in an area that has become known as 'White Shark Café', about 2,000km (1,243mi) from the American coast and half way to Hawaii. During the journey, they swim close to the surface, rarely diving deep, but when they reach the 'café' they dive deeper, down to 500m (1,640ft) and back again, day and night, making 150 vertical excursions within twenty-four hours. Males dive more frequently than females. Are they feeding or are they breeding? The behaviour, researchers suggest, could be interpreted in either of two ways.

Firstly, the 'café' could be just that – a feeding area, where sharks are diving down to intercept the daily vertical migration of ocean animals, remaining close to the surface at night and going deeper during the day. Secondly, it could be a mating area where, in diving, the males are either showing their prowess to the onlooking females, like birds at a lek, or they are searching for available females. In this scenario, the females would spend less time at the centre of the 'café' area, because mating in sharks is a rather rough affair; females would rather slip in, mate and get out in a hurry.

However, some sharks go all the way to Hawaii, and they stick around there. Another study has revealed that males are present from December through to June and females are present every month except November; the reason why is not clear, although there is plenty of food to attract them – winter aggregations of humpback whales *Megaptera novaeangliae* with newborn calves and discarded placentas readily available, along with spinner dolphins *Stenella longirostris*, which congregate each day in Hawaii's bays, and monk seals *Monachus schauinslandi*, which are present all year round. All the great whites seen in these waters have been mature sharks

3.3–4.5m (11–15ft) in length; and in a third study there is speculation that pregnant female sharks, rather than mating in the open ocean, hang out there and segregate from males, especially in warmer waters, which speeds up foetal development, until they are ready to give birth.

This new research followed great whites in the eastern Pacific over a period of years rather than months and it has revealed that mature female great white sharks in the eastern Pacific have a migration pattern based on a two-year cycle. The sharks in this study aggregate off Guadalupe Island on the Mexican coast for part of their inshore phase and travel to the 'café' when offshore. While males migrate annually between the two areas, females remain out in the ocean for fifteen to sixteen months, the best part of the gestation period for this species; so it's thought these are pregnant females that segregate from the males in the open ocean rather than meeting with them, before heading back to the Mexican coast to drop their pups at two separate pupping sites. This is followed by a visit to Guadalupe Island where they not only feed on seals and sea lions but also meet with males and mate. Duty done and, no doubt, a haunch of sea lion in their bellies, they head back out to the open sea.

OCEAN'S DUSTBIN

The notorious tiger shark *Galeocerdo cuvier* is known as the 'ocean's dustbin' for it eats just about anything. It also shares with the great white shark the title of 'man-eater' on account of it biting people and, like the great white, it can grow to immense size. It is up there with the world's largest predatory sharks, with lengths of 5.5m (18ft) reliably measured. A female tiger shark caught somewhere off the coast of Indo-China in 1957 was reported to be 7.4m (24ft) long, and monster sharks exceeding 9.1m (30ft) have been claimed but not confirmed.

Tiger shark

In the recent past, when tiger sharks have bitten people in Hawaii, the authorities have slaughtered every shark in the vicinity of the attack site in a desire to eliminate the 'rogue shark'. The thinking was that tiger sharks were attached to a particular local home range and so did not go far. However, recent research by the Hawaii Institute of Marine Biology has confirmed that after such an incident any shark would have long gone. A shark might travel 35km (22mi) or more, even across deep-water channels, to visit other parts of the archipelago, so a culling programme was proved to be totally ineffective. Some sharks even head out to the most westerly islands, such as the French Frigate Shoals, where they turn up at exactly the same time each year, just as Laysan albatross *Phoebastria immutabilis* chicks are fledging. The young birds are not especially good flyers and some ditch in the sea. The tiger sharks wait below the waves ready to catch them. How they know when to arrive at the shoals is another mystery.

SNIFF 'N' SNACK

The French underwater explorer Jacque-Yves Cousteau considered the oceanic whitetip shark *Carcharhinus longimanus* to be the most dangerous shark in the sea. It is known to be an ocean wanderer, and is often the first species of shark to pitch up at shipwrecks and harass people in the water. Seemingly fearless, oceanic whitetips do not circle their prey warily, like many other sharks, but head straight in, bumping the target to check for palatability before taking the first bite. They are instantly recognisable by their spade-like fins (something of an Achilles heel as they are valued in the shark fin soup industry), and are generally encountered in warm seas where the water temperature is above 18°C (64°F), although one stray ended up on a beach on the west coast of Sweden. How and why it came to be there is puzzling, to say the very least.

The oceanic whitetip is not high in the shark bite incident league tables, with just a handful of recorded attacks on people, but its presence at shipwreck disasters, such as that of the USS *Indianapolis,* sunk between Guam and Leyte Gulf on 30 July 1945, and the troopship *Nova*

Scotia, sunk close to the South Africa coast on 28 November 1942, means that many shark attacks by this species have probably gone unrecorded. In more recent times, it was thought to be responsible for several attacks on tourists in the Red Sea, near Egypt's Sharm el-Sheikh in November and December 2010. Apparently an individual, recognised by a bite mark on the upper lobe of its tail fin and the sparse pattern of white on its dorsal fin, had been hand-fed by divers and began to associate people with a free meal. It targeted the buttocks and thigh, which is where divers and snorkellers tend to keep their catch strapped to their dive belt.

Oceanic whitetip shark

Until now, it was thought that oceanic whitetips roam the oceans with very little pattern to their movements, but research at Stony Brook University in New York has revealed that female oceanic whitetip sharks found around Cat Island in the Bahamas tend to remain there for much of the year. Six of the eleven sharks fitted with satellite tags patrolled the Bahamas exclusive economic zone, an area about the size of France. They dived for short periods of about fifteen minutes to forage down to 1,000m (3,281ft); the temperature was significantly lower than at the surface so they descended slowly and ascended rapidly in order to warm up before their next hunting period. The other five stayed within 500km (311mi) of Cat Island for up to thirty days before fanning out across 16,422 sq. km (6,341 sq. mi) of the western North Atlantic, and then returned to the Bahamas up to 150 days later. It is thought this travel schedule, like that of female great white sharks, is linked to their two-year reproductive cycle. They head off to mating sites or birthing grounds before returning to their 'home' in the Bahamas. Just as great whites were always thought of as stay-at-home creatures and are turning out to be ocean wanderers, oceanic whitetips were thought to be ocean wanderers and are turning out to be home-lovers. It just shows how little we know about the animals of the sea.

Sharks have a battery of senses with which to find their prey. They can hear distant sounds and vibrations, detect movements in the ocean currents, see well and sometimes in colour. They can also detect the electrical activity of their prey's muscles, and smell minute quantities of body fluids, such as blood, in the vastness of the ocean and follow an olfactory gradient back to its source, such as a bleeding fish or marine mammal. This last sense, however, has thrown up something of a conundrum. Sharks, such as the oceanic whitetip shark which lives in the open ocean and investigates anything that might constitute a meal, appear much quicker than you would expect. Chemicals disperse in water relatively slowly compared to those in the air, yet the sharks sometimes pitch up within minutes of a release. How do they do it? It seems they might not rely on the smell travelling through water, but can pick up odours that are blown on the wind above and across the sea's surface. These open-ocean sharks tend to have their nostrils higher up on the snout than, say, dogfish that live close to the seabed, so researchers at the Research Institute for Human Morphology in Moscow have suggested that they can push their snouts above the water and capture air bubbles containing the smell, which are then retained in the nostrils. The shark nose can only work with scents dissolved in water, but the oceanic whitetip has many folds in its nostrils that would work to break up the bubbles and release the scents, which are then detected by the chemosensory cells lining the nasal cavity. Some surface-dwelling sharks, such as the great white shark, are known to lift their head out of the water to look around, so there's every chance that these types of sharks could detect airborne scents in a similar way to land scavengers.

THE ADAPTABLE BULL SHARK

Bull sharks *Carcharhinus leucas* are the bruisers in the shark world. They sneak up on their prey in murky waters, such as estuaries and harbours, and will even go into rivers. They have a bump-and-bite feeding strategy and their propensity to hold territories in shallow waters brings them into contact with people. They grow to about 3.5m (11.5ft) long, females being larger than males.

Size for size bull sharks have the strongest bite of the thirteen sharks, including great white sharks and great hammerheads, tested at the University of South Florida at Tampa. Young bull sharks bite harder than larger, older individuals – a possible adaptation to enable them to feed on a greater variety of prey than that targeted by adults.

Bull shark

The bull shark is one of the few sharks that enter freshwater, which has given rise to some of its other common English names, the Zambezi shark or Lake Nicaragua shark, after the bodies of water in which it is found. The sharks tend to be females, about to give birth in the safety of the river or lake, or youngsters growing up and on their way out. They can tolerate freshwater because, unlike most other sharks, they have the wherewithal to maintain the correct salt and water balance. At one time the Lake Nicaragua sharks were thought to be a separate, entirely freshwater species, but now we know that they enter and leave the lake via the San Juan River, leaping up the rapids like salmon. Many rivers play host to these sharks. Bull sharks have been seen as far inland as Kentucky on the Ohio River and Alton, Illinois, on the Mississippi. They also appear in Maryland's Potomac River, the Ganges and Brahmaputra rivers in India and Bangladesh, and a population of about 500 is to be found in the Brisbane River, Queensland. They've even swum 4,000km (2,485mi) up the Amazon, and have attacked hippos, crocodiles and cattle 200km (124mi) up the mighty Zambezi. During the Australian floods of 2010–11 bull sharks were seen swimming down streets in Brisbane and Goodna, Queensland, and golfers at the fourteenth tee of the Carbrook Golf Club in Brisbane regularly see their dorsal fins on the surface of a large lake in the middle of the course; the sharks were trapped there after a flood and now the golfers hold a monthly

'Shark Lake Challenge'. Children used to jump into the lake to recover stray golf balls but they certainly don't do that any more.

SHARKS THAT BITE PEOPLE

Three sharks dominate the serious shark bite statistics.

Great white shark *Carcharodon carcharias*

The great white is responsible for the largest number of severe attacks on people. It patrols seal and sea lion colonies on the lookout for unsuspecting prey, so biting people could be cases of mistaken identity. People look like exceptionally slow and clumsy seals in the water, and predators are programmed to seek out the weak and the helpless. However, many people survive being bitten by an animal that could swallow them whole, probably because they just don't have enough fat on them to make feeding worthwhile or simply because the shark is wary of creatures with which it is unfamiliar. It is, perhaps, significant that people diving outside a cage with great whites are rarely approached as long as they have eye contact with the shark. This species relies on a surprise attack from below and behind.

Tiger shark *Galeocerdo cuvier*

Tiger sharks eat just about anything so people must be considered fair game. Even so, attacks are not common.

Thousands of people bathe, surf and dive around the Hawaiian Islands every day of the year, yet during 2012 there were only ten unprovoked shark bite incidents, most of which were probably attributable to tiger sharks. Again, these could be cases of mistaken identity. People on bodyboards look very similar to sea turtles when viewed from below, and turtles are a favourite tiger shark food. Tiger sharks also frequent shallow reefs and waters close to shore so they are more likely to come into contact with people.

Bull shark *Carcharhinus leucas*

It is thought that shark bites attributed to other sharks, such as great whites, could actually have been by bull sharks. In 1916, for example, a series of fatal attacks along the New Jersey shore, which were thought to be an inspiration for Peter Benchley's *Jaws*, were at first attributed to a great white shark, but the actual culprit was probably a bull shark. Similarly, attacks in Sydney Harbour, originally pinned on great whites, were probably by bull sharks.

Despite the headlines and the horror associated with a shark bite incident, mercifully they are extremely rare. According to the International Shark Attack Files (ISAF) held at the Florida Museum of Natural History (University of Florida), far more people are drowned at the beach than are killed by sharks. Taking the ten years beginning

1990, fifteen to twenty-five people drowned each year on California's beaches while only one person died from being bitten by a shark during the entire decade.

To put the frequency of shark attacks further into perspective, from 2001 to 2010 there were 263 fatalities from domestic dog attacks in the USA (according to the National Canine Research Foundation), while the ISAF reports that there were no more than ten fatal shark attacks in the USA during the same period.

Blacktip shark

Other types of sharks have been implicated in minor incidents. Two notables are the blacktip shark *C. limbatus* and the blacktip reef shark *C. melanopterus*. The blacktip shark, which is sometimes seen migrating in huge numbers along the Florida coast, accounts for 16 per cent of shark bites off Florida beaches. It's thought the shark mistakes white skin, especially on the palms of hands and soles of feet, as the flash of fish scales in the surf and is just as surprised as its victim when it seizes a leg attached to a rather large creature and not a more manageable fish. The bite victim may need stitches from the gash, but it is rarely life-threatening. Similarly the blacktip reef shark bites legs, feet and ankles of people wading in shallow water on coral reefs.

Blacktip reef shark

It has been known for some time that the thing that sharks detest most is the smell of dead sharks, but until now scientific studies on the subject have been inconclusive. During the past five years, however, Shark Defense of New Jersey has initiated renewed interest. Their scientists squirted extract of rotting shark from aerosol cans into feeding frenzies of blacknose sharks *C. acronotus* and Caribbean reef sharks *C. perezi* off the coast of Bimini in the Bahamas and watched what happened. Just 150ml (0.3 pints) caused the sharks to scatter within a minute and they stayed away for a further ten minutes or more. The team tested bubbles, aerosol propellant, sound and several other relevant chemicals but it was only essence of rotting shark that stopped the feeding activity.

DEADLY JELLIES

The world's most dangerous jellyfish are undoubtedly those in the box jellyfish group, especially Australia's box jellyfish or sea wasp *Chironex fleckeri*, the largest box jellyfish and one of the most venomous animals in the world (see page 125). Its venom is said to be 100 times more potent than that of the cobra; it can cause the heart to stop and death to occur in less than five minutes. The equally potent Japanese Habu-kurage *Chironex yamaguchii* has caused human deaths along the Japanese coast and in the Philippines; the four-handed box jellyfish *Chiropsalmus quadrumanus* has killed in the Gulf of Mexico; and the tiny, thumbnail-sized Irukandji jellyfish *Carukia barnesi* and *Malo kingi*, which have stingers on the bell as well as the four long tentacles, have been responsible for deaths in Australia. All these jellyfish are recognised by their transparent cube-shaped bells with which they can actively swim much faster than the more conventional jellyfish with their saucer-shaped or hemispherical bells. Box jellies can swim in a particular direction, at a speed said to be comparable to an Olympic swimmer, and can avoid obstacles for they have twenty-four true eyes, each with a retina, cornea and lens, similar to human eyes, which are placed on the four sides of the bell. The most dangerous species lives mainly in the Indo-Pacific region, although one species of the Irukandji jellyfish has been found off the British Isles and on the Florida coast. People such as beach lifeguards, who must work in areas where box jellyfish are common, wear tights for protection. It was first thought that this prevented the barbs from the sting cells from reaching the skin, but more recent research indicates that the cells are triggered to fire by chemicals on the skin and the tights eliminate this stimulus.

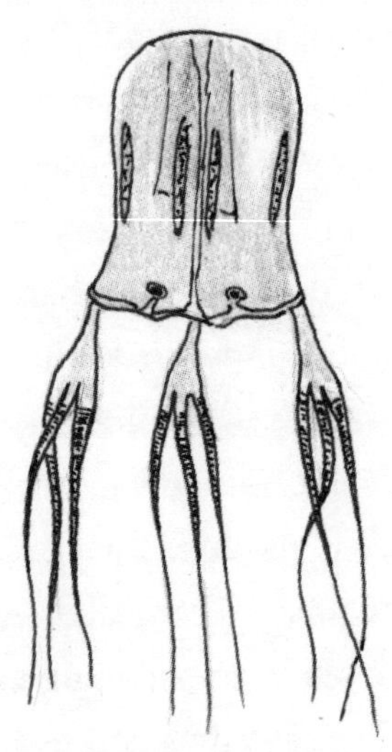

Box jellyfish

In November 2007 a huge swarm of mauve stinger jellyfish *Pelagia noctiluca*, estimated to be 16km (10mi) across and 13m (43ft) deep, drifted in to the County Antrim coast of Northern Ireland and wiped out most of the stock – over 100,000 fish worth over one million pounds – at the country's only salmon farm. There were so many jellyfish the sea was turned red.

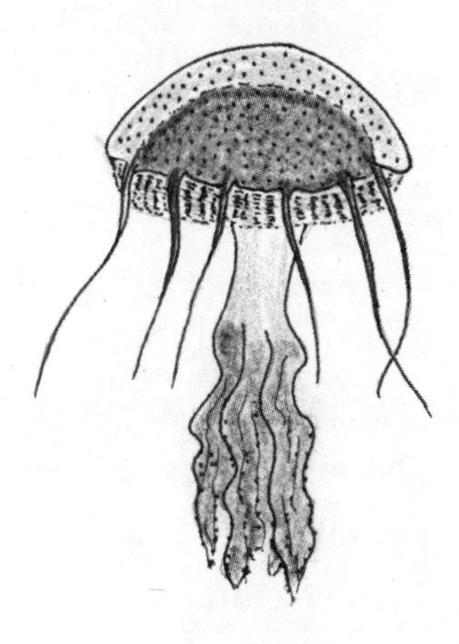

Mauve stinger jellyfish

The largest jellyfish in the sea is the Arctic giant jellyfish *Cyanea capillata arctica*, an outsize version of the lion's mane jellyfish that is frequently found along the shores of the British Isles. The Arctic subspecies, which lives in shallow waters in the north-west Atlantic, can take on monster proportions. In 1865, a specimen observed and measured in Massachusetts Bay, near Boston, had a bell 2.29m (7.5ft) across and tentacles that stretched 36.6m (120ft) below, making it one of the longest animals in the world, longer even than a blue whale.

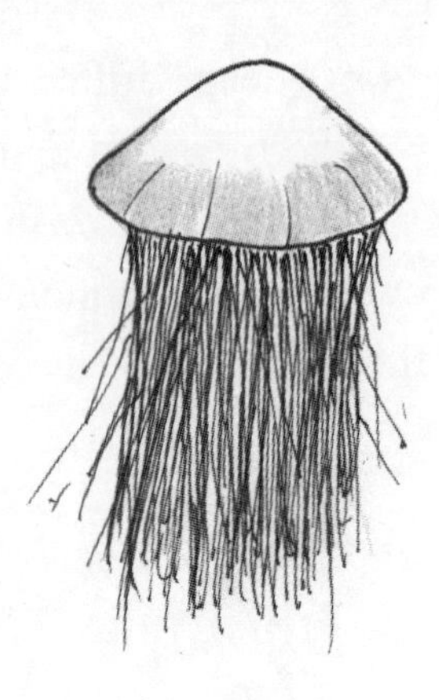

Arctic giant jellyfish

It was always thought that the majority of jellyfish were at the mercy of the winds, tides and ocean currents, capable only of moving up and down in the water column and little else. Then, researchers from the University of California at Los Angeles looked at common moon jellyfish *Aurelia aurita* in Saanich Inlet, a 22km (14mi) long fjord at the southern end of Vancouver Island. They found that in early summer all the jellyfish are at the northern seaward end of the inlet, but by September they have gathered in two swarms at the south-east end. They also noticed that the jellyfish have two different swimming positions. They saw the familiar vertical movement achieved when the bell faces up or down, but there is also horizontal movement with the bell moving parallel to the surface. At night the jellyfish rise up and down or have no movement at all, but at sunrise they adopt the horizontal swimming mode and head towards the sun. If a cloud covers the sun or if they are in the shadow of a mountain, they move randomly, but if the cloud passes and the sun comes out they immediately resume their horizontal movement until nightfall. It seems these relatively simple creatures use the sun as a compass.

Moon jellyfish

In a saltwater lake on Eil Malk island in Palau, in the western Pacific, there are huge swarms of golden jellyfish *Mastigias* cf. *papua etpisoni*. They have been isolated here for over 12,000 years and they too follow the sun, from one side of their lake to the other, for they are totally dependent for food on tiny algal cells living in the tissues of their ruff-like tentacles. The algae need the sun for photosynthesis so the jellyfish rise up in the morning, track it across the lake and then sink back down in the evening to a level where they can acquire nitrogen for their symbiotic algae, only to start the cycle all over again the following day. They also revolve counter-clockwise so that all their algae get a chance in the sun. However, their movement across the lake might also be driven in part for another reason. Jellyfish-eating sea anemones *Entacmaea medusivora* reside in the eastern part of the lake, and by staying in the sun and avoiding shadows the jellyfish tend to stay clear of the anemones, leaving the population of moon jellies *Aurelia*, which share the lake but do not migrate across it, to satisfy the anemones' appetites.

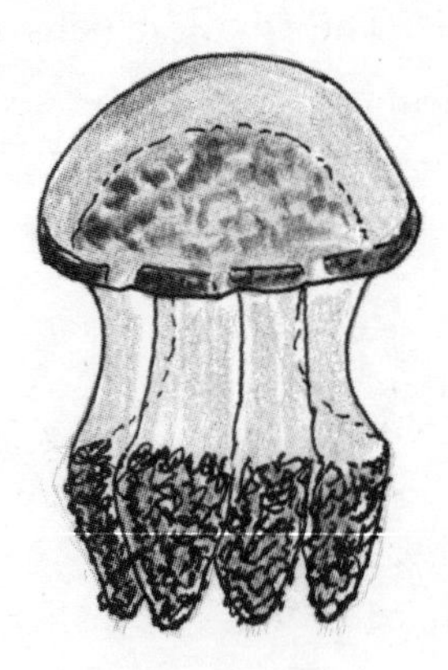

Golden jellyfish

In November 2009, giant jellyfish sank the 10-ton Japanese fishing trawler *Diasan Shinsho-maru*. The jellyfish were another of the world's largest species, known as Nomura's jellyfish *Nemopilema nomurai*, each of which can weigh up to 200kg (441lb) and grow to 2m (6.6ft) across. When the fishermen hauled in their net the weight of the mass of jellyfish they had caught caused their ship to capsize and they were thrown into the water, from where the crew of another trawler rescued them. The jellyfish appear periodically in Japanese waters. Some years none are seen, while in others they appear in their thousands, causing considerable damage to fishing gear and even the inlet pipes to power stations.

The world's smallest jellyfish are what are called 'creeping jellyfish' *Staurocladia*, for they use their tentacles to adhere to and move about on seaweeds, where they feed on copepods. The smallest bells are about 0.5mm (0.02in.) across, and can only be seen clearly with a microscope.

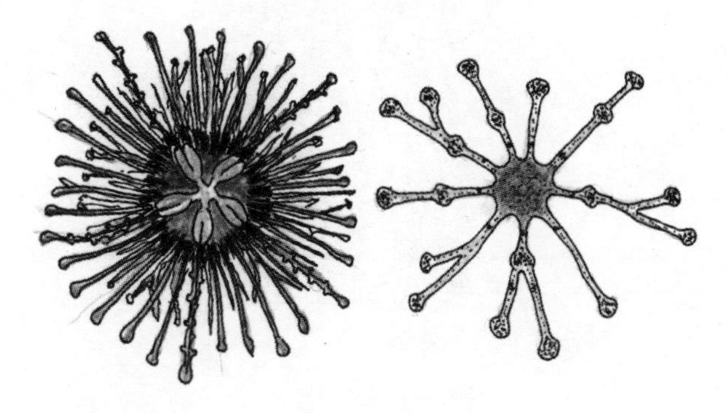

Creeping jellyfish

The Portuguese man o' war *Physalia physalis* is another dangerous stinging machine. Technically it is not a jellyfish, which is a single organism, but a siphonophore, made up of many organisms known as zooids, each with its own job to do. Some form the tentacles – typically about 10m (33ft) long – which catch food, and others digest it, and still others specialise in reproduction. A fourth type forms the gas-filled float, which catches the wind like a sail and enables the creature to move independently of the ocean currents. The sting cells inject potent venom, which has been known to kill humans – especially people with weak hearts or who have a severe allergic reaction to the venom – and on the east coast of Australia alone, on average 10,000 bathers are stung each year.

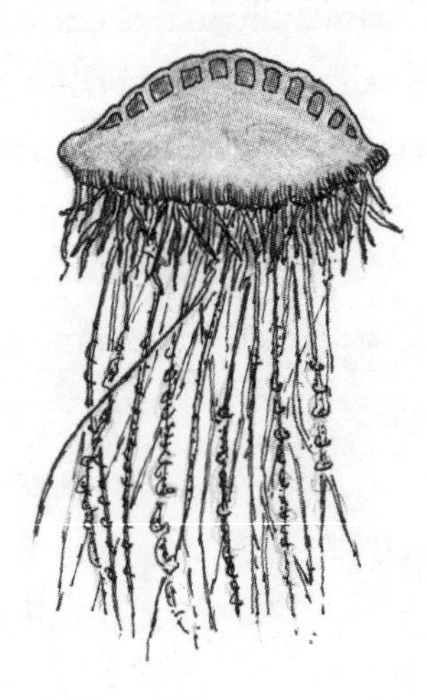

Portuguese man o' war

THE OCEAN'S MOST VENOMOUS ANIMALS

Box jellyfish (Class: Cubozoa) are considered the most dangerous animals on the planet. Stings are extremely painful and can result in death from heart failure in as little as two minutes. The most dangerous species lives in the tropical Indo-Pacific region, where deaths are linked to the lack of readily available first aid. In the Philippines, for example, there are up to forty deaths a year.

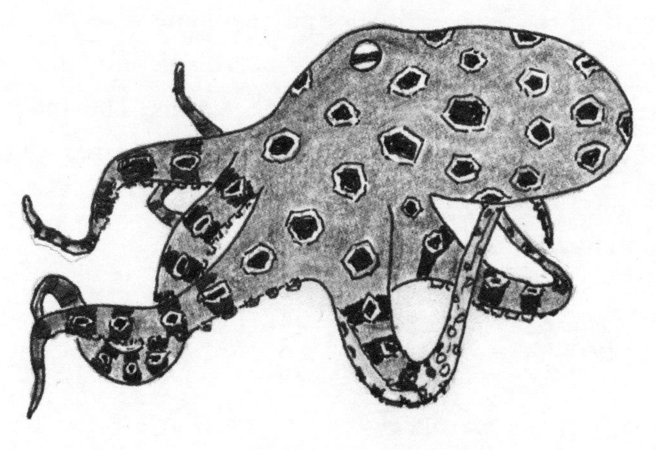

Blue-ringed octopus

The blue-ringed octopus *Hapalochlaena* is small and docile, but when riled the spots on its body become more pronounced and the blue rings pulsate; then, it is ready to bite. Its venom, which is manufactured by bacteria lodging in its salivary glands, can kill humans by cutting off the supply of oxygen to the brain. There is no antivenom and no known antidote to the powerful neurotoxin, known as tetrodotoxin, more usually found as a poison in the skin of toads. This is the first time it has been found in a venom. There are three species that live in the Indian and Pacific oceans, from Japan to South Australia.

The faint-banded sea snake *Hydrophis belcheri* delivers the most potent venom of all the sea snakes, and is third in the league table of all venomous snakes, after the inland taipan *Oxyuranus microlepidotus* and the eastern brown snake *Pseudonaja textilis*. The species, however, is mild-mannered and rarely bites. It is found in the tropical Indian Ocean region as far east as the Solomon Islands.

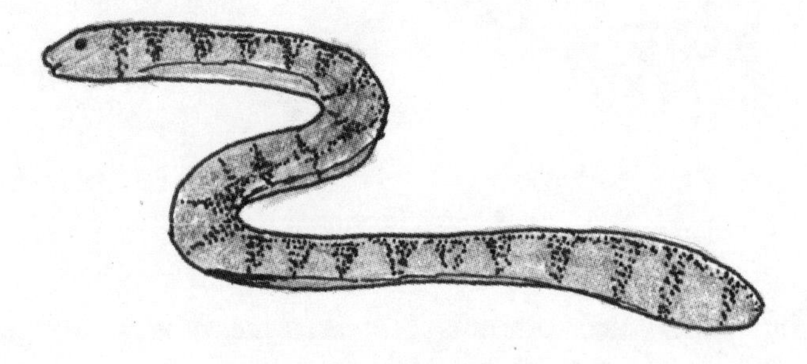

Faint-banded sea snake

The large cone shells *Conus*, which catch fish rather than worms, are the most dangerous types. The geography cone *C. geographus*, which is found on Indo-Pacific coral reefs, delivers the most toxic venom from a harpoon-like tooth thrust by an extendable proboscis. There is no antivenom. The venom, however, is an effective pain-killer, 10,000 times more potent than morphine and not addictive, and it can target human pain receptors – medicine from nature.

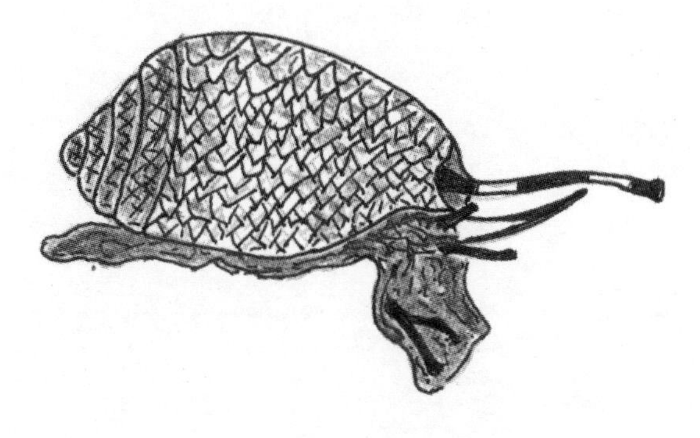

Cone shell

The reef stonefish *Synanceia verrucosa* is probably the world's most venomous fish. It is also one of the best camouflaged. Stonefish rest on the seabed, among the corals, and look just like stones. Bathers inadvertently stand on them. Hollow dorsal fin spines deliver the venom, and they can even penetrate thick-soled boots. The pain is excruciating; victims are said to demand that the affected limb be amputated. There are five closely related species and they are found on reefs in the Indian and Pacific oceans, as well as the Caribbean and around Florida.

Reef stonefish

TALES OF WHALES

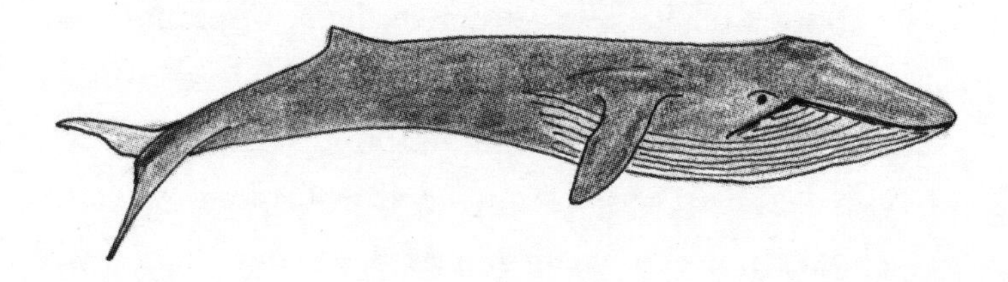

Blue whale

The blue whale *Balaenoptera musculus* is not only the world's largest living animal, but is also thought to be the largest animal that has ever lived. The biggest reliably measured specimen was landed at the shore station of the Compañía Argentina de Pesca (Argentine Fishing Company) at Grytviken on South Georgia during the heyday of Antarctic whaling, sometime between 1904, when the station opened, and 1920. It measured 33.57m long (110.14ft). Many more whales over 30m (98ft) were harpooned during those whaling years in the Southern Ocean, but today blue whales of that size are exceedingly rare. Whales are difficult, if not impossible, to weigh whole, so an accurate weight is not easy to gauge; by weighing it piecemeal, blood and body fluids are lost. Even so, estimates of 180 tonnes have been claimed. Its body parts also throw up some staggering statistics: the tongue can weigh as much as an elephant and the heart is almost as large as a small car – and all this on a diet of tiny shrimp-like krill and copepods, 40 million krill per day.

How about a whale with legs? Ex-whaler Captain Dode MacPhearson recalls a blue whale (known also as a 'sulphur-bottom whale') being harpooned that had very obvious legs. The story is included in Captain William Hagelund's book *Whalers No More*, about the British Columbia whaling industry (1910–41), and it tells of an occasion when whaler Captain Willis Balcom had difficulty lashing the dead whale alongside his whale catcher because of the protuberances. Unfortunately, Willis hacked off one of the legs, much to the consternation of zoologists who were intrigued by the malformation, but proof that such a thing did occur exists in the form of some of the whale's bones, which were preserved and are apparently on show at a local museum.

American naturalist Roy Chapman Andrews called the fin whale or finback *B. physalus* the 'greyhound of the seas'. Its streamlined body is the second longest of living whales, with a maximum length of 27.3m (89.5ft), but its shape and powerful swimming muscles ensure that it is one of the fastest baleen whales. Sustained speeds of 37–41km/h (23–26mph) have been recorded, along with bursts of

46km/h (29mph). It is also the most gregarious of whales and may roam the oceans in loose herds while remaining in contact with very-low-frequency contact calls, a behaviour known as 'counter-calling' during which one whale in the pod calls to another and the other responds. It is thought that the quality of the incoming call gives the whale information not only about the caller, but also the echoes contain information about their surroundings. One of these calls was a mysterious single '20 hertz sound' that was thought to be from submarines. It was not until scientists placed a three-dimensional hydrophone array on the seabed that they realised the calls were coming from fin whales widely spaced apart but within 'shouting' distance. Males also make very loud patterned low-frequency calls, between 15 and 30 hertz, during courtship. They can last for fifteen minutes at a time and be made for several days.

Fin whale

The ocean's greatest virtuoso must be the humpback whale *Megaptera novaeangliae*. On the breeding grounds the male humpback sings extraordinarily long and sonorous songs, which are true songs like those of birds, with repeated phrases. All the whales in a part of the ocean sing the same song, which evolves through time, so whales, say, in the Indian Ocean sing very different songs from those in the northern Pacific. Despite many years of intense research the function of the songs is still unknown. Attracting females, male competition and the establishment of dominance, and males summoning allies in the fight for females have all been suggested at one time or another, but recent research shows that singers often (68 per cent of the time) associate with a mother and calf. 'Why' is still the question waiting to be answered. The whale itself is easy to identify: it has the longest pectoral fins of any whale. Bumps or tubercles along the leading edge of the pectorals are thought to give it greater hydrodynamic efficiency and researchers are now copying the design for future wind turbine blades.

Humpback whale

A female humpback whale holds the world record for a mammal on migration. She travelled at least 9,800km (6,089mi) from breeding grounds on Abrolhos Bank, off the Brazilian coast, across the Atlantic to breeding sites close to the east coast of Madagascar in the Indian Ocean. This is about twice the average distance covered during the annual migration of this species between feeding grounds in polar seas and breeding sites in the tropics, and the longest documented journey of any mammal. She beat the previous record holder – a humpback whale that travelled between American Samoa and the Antarctic Peninsula – by 400km (249mi).

In the Gulf of Maine humpback whales commonly feed by blowing bubbles below a school of sand lance *Ammodytes hexapterus* and then lunging through the bubbles with mouths agape. In 1980, however, one inventive individual adopted a new technique. It slapped the sea's surface with its tail flukes and then blew bubbles, a behaviour that became known as 'lobtail feeding'. During the next couple of decades gradually many more whales copied the feeding behaviour, an example of cultural transmission, something humpbacks share with humans and non-human primates. There are innovators in the whale population, just as there are in human populations. They come up with new ways to do things and their ideas gradually spread to the other whales. Humpback song (see above) evolves in the same way, with a single virtuoso changing the song and every other male whale in the population copying the change. Whales, like people it seems, have multiple traditions.

Bowhead whales *Balaena mysticetus* live in Arctic and sub-Arctic waters where they feed by skimming the water with half-open mouths, their bristle-like baleen plates sieving the water and trapping tiny organisms, such as copepods. It was always thought that baleen, which is made of the same substance as hair and fingernails, was a passive structure, but experiments at Hampden-Sydney College in Virginia revealed that at the bowhead's normal swimming speed, when it is 'ram-feeding', the bristles fan out to form a dense, tangled net that traps considerably more food than at slower or faster speeds.

Bowhead whale

In fact, bowheads are turning out to be quite exceptional whales. They not only have the world's biggest mouth, but they seem also to have an extraordinary way to keep cool. On the roof of a whale's mouth is an organ that resembles very closely the engorged male penis and it helps to keep them cool. The scientists from Hampden-Sydney College have examined dead bowheads caught legally by Alaskan hunters and have found that a rod of tissue, which they have called the *corpus cavernosum maxillaris*, runs along the middle of the palate; they speculate that this becomes swollen with blood and serves to let off excess body heat. Evidence rests with the fact that the organ remains at a temperature of 6–8°C (10.8–14.4°F) warmer than other tissues even several hours after the whale has died, indicating that it effectively serves to transfer heat from the warm body of the whale into the cold seawater. Bowheads are well insulated from the icy Arctic waters, but in summer when they are on migration their bodies could overheat from the exertion, so they simply open their mouths and let the cool water swill around their blood-filled, penis-like mouth organ – the world's biggest erectile secret.

The grey whale *Eschrichtius robustus* once had the worst reputation of all the baleen whales. It was known as the 'devil fish' because defensive mothers with calves would ram whaling boats on the calving lagoons in Mexico. Today it is a gentle giant, allowing eco-tourists to come alongside and stroke its snout. It is one of the species of whales that embark on extensive annual migrations from Mexico to Alaska and back. On its feeding grounds in the Arctic it mainly feeds on shrimp-like amphipods that it scoops from the mud on the ocean floor, making it the only baleen whale to feed on bottom-dwelling organisms. The whales do this by turning on their sides; so scientists can tell whether a whale is right-handed or left-handed by looking at the lack of barnacles on the side that scrapes along the seabed.

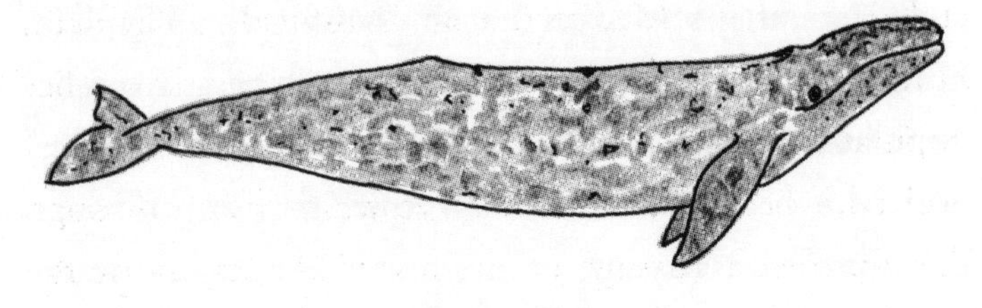

Grey whale

In May 2010, scientists with the Israel Marine Mammal Research and Assistance Center spent about two hours watching a 12m (39ft) long whale behaving as whales normally do. It was swimming close to the surface, making short dives and occasionally showing its tail flukes. At first the researchers thought they were following a sperm whale, a species that is seen in the Mediterranean but which is a rarity off the Israeli coast. However, as they were making their observations, they noticed that the blowhole was not at the front of the head as on a sperm whale but further back like one of the baleen whales. Examining their pictures back at base they discovered that they had been watching a grey whale. The only problem is that there have been no grey whales in the Mediterranean, or in the entire Atlantic come to that, since the eighteenth century, when whalers hunted the Atlantic stock to extinction. Today, there are two populations surviving, both in the Pacific Ocean: one that migrates between Baja California and

Alaska in the eastern Pacific and another found along the Kamchatka coast in the west. So, how a grey whale ended up in the Mediterranean is difficult to explain. However, there are two possible scenarios: firstly, a relict population has survived in the Atlantic but no one has noticed it before, or secondly, the whale swam through the Northwest Passage to the north of Canada or the Northeast Passage to the north of Eurasia, aided by the current lack of ice in the Arctic in summer, and then headed south in the Atlantic, before turning left into the Mediterranean (possibly following the same pattern as its normal autumn journey when it heads south in the Pacific and takes a left into the Gulf of California). The whale was seen twenty-two days later off the coast of Spain, but what happened to it after those observations is not known, as the researchers had no funding to satellite tag the whale. However, on 4 May 2013, another one turned up at Walvis Bay on the coast of Namibia – the first time a grey whale has been seen in the Southern Hemisphere. It is unlikely to have been the same whale, but it is thought that it must have entered the Atlantic in a similar way to the Mediterranean whale.

Bryde's whale (pronounced Brew-da) *Balaenoptera edeni* lives permanently in tropical and temperate seas. It is one of the whales most frequently found following the 'sardine run' off the South African coast (see page 1). It barrels up from the depths and tries to engulf any bait ball that has been corralled by dolphins – but it is slow to manoeuvre and many fish get away. It then has to laboriously turn around, like a super-tanker turning, and try again, by which time the dolphins, sharks and gannets have vacuumed up most of the rest of the fish. The longest individual was measured at 15.51m (50.9ft), putting this whale between the slightly larger sei whale *B. borealis* (the third largest species of living whales) and the much smaller minke whales *B. acutorostrata* and *B. bonaerensis* in size.

While people go 'whale watching' in many parts of the world, off Australia's Great Barrier Reef dwarf minke whales (a subspecies of *B. acutorostra* or maybe a species in its own right, but yet to be named) go 'people watching'. Tourists in wet suits and snorkels hold on to a line extending from a boat and wait, for it is against Australian law to approach a whale; the whale must initiate the encounter. Gradually the whales, each 7–9m (23–30ft) long, arrive and approach the divers, their white dorsal fins clearly visible even in the gloom. Then they nod their heads up and down while calling quietly at frequencies within the human hearing range, and literally look each person in the eye. It's an altogether once-in-a-lifetime experience.

TOOTHED WHALES

Sperm whales *Physeter macrocephalus*, which grow to 20m (66ft) long, may be the largest hunters in the sea, but they come under attack themselves from significantly smaller predators, such as orcas or killer whales *Orcinus orca* that work together, like wolves and wild dogs on land, to subdue larger prey. Calves, juveniles, the sick and infirm are most at risk, especially those in female herds, bulls being far too big for orcas to tackle. As a defence the herd forms a circle with the most vulnerable at its centre. The adult females gather together with their heads pointing towards the centre and their powerful tail flukes facing outwards, poised to repel the attackers – the so-called 'Marguerite formation', named after a flower, the Marguerite daisy *Argyranthemum frutescens*. Sometimes they reverse their position. With their heads facing outward, they use their jaws and teeth to fend off the attack.

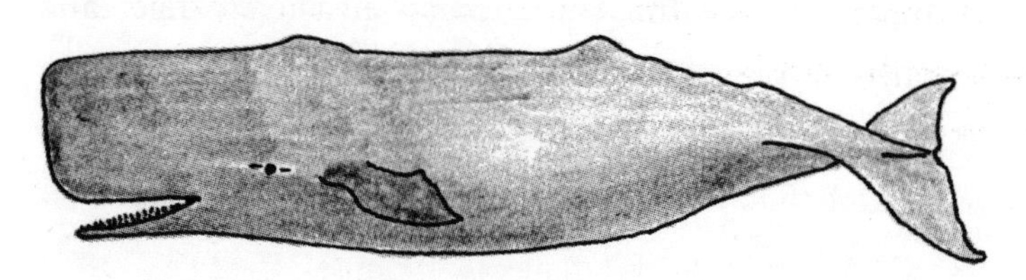

Sperm whale

Sperm whales suffer an irritation in the stomach caused by the sharp beaks of squid and octopuses. To deal with this they produce an evil-smelling substance called 'ambergris', which is almost worth its weight in gold, for it is used by the perfume industry to make fragrances in perfumes last longer. Sometimes the lumps are regurgitated and float away in the ocean, only to be washed up on the shore where a lucky beachcomber might find them. The largest known piece weighed 455kg (1,003lb) and was found by whalers in Norway in 1908. Other sizeable 'boulders' of ambergris include a 411kg (906lb) chunk in New Zealand in 1883, a 420kg (926lb) piece found in the Antarctic in 1953 and a 138kg (304lb) lump discovered on the Falkland Islands in 1927. More recently a couple on a beach in Australia found a chunk weighing 15kg (33lb), which netted them 295,000 Australian dollars; and in January 2013 a man walking his dog on Morecambe beach on the west coast of Britain found a piece thought to be worth about £100,000.

Two whales – a cow and a calf – washed up on Opape beach in New Zealand's Bay of Plenty in December 2010 turned out to be specimens of one of the rarest and least observed species of whales, the spade-toothed beaked whale *Mesoplodon traversii*. In fact, the species is so rare that until these two appeared scientists were unsure whether it still existed at all. It is known only from jawbones and skulls and has never been seen alive, for it normally lives in the deep sea, surfacing only briefly to breathe. At first scientists thought the beached whales, the larger of which was 5.3m (17.4ft) long, were the more common Gray's beaked whale *Mesoplodon grayi*, but DNA analysis revealed it to be the less well-known species, characterised by overly large teeth, up to 23cm (9in.) long, which resemble flensing spades, one of the tools used by whalers to hack the blubber off whales – hence the common English name. It is thought that normally their dead bodies simply drop into the depths of the sea, so why they came to be washed up on a beach is not known.

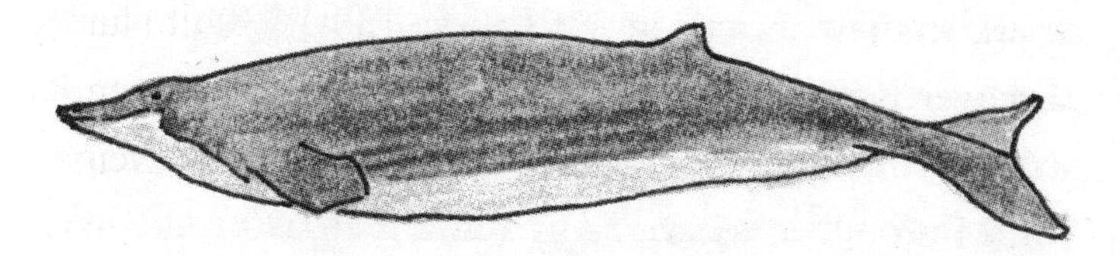

Spade-toothed beaked whale

Short-finned pilot whales *Globicephala macrorhynchus* have been seen to carry the dead bodies of California sea lions *Zalophus californianus* that have been shot by fishermen. The bodies are in various stages of decomposition; however, the pilot whale neither eats them nor plays with them, but pushes them along with its snout or drapes them over its dorsal fins and even uses its flukes to tow them along. It will swim with a corpse in its mouth and dive down with it; researchers are at a loss to explain why. One theory is that it shows the strength and therefore the fitness of an individual, and so the behaviour is an indicator of the carrier's social status.

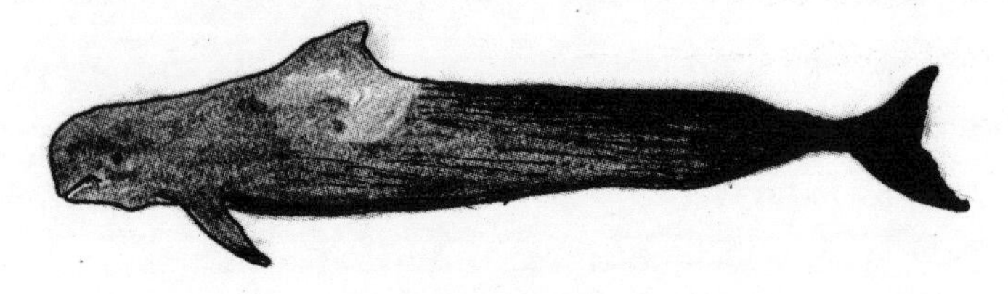

Shortfin pilot whale

The narwhal *Monodon monoceros* is an extraordinary animal. The male (and sometimes the female) has an exceptionally long, helical tooth that erupts from the lower left jaw. It can be up to 3m (10ft) long, and its structure is unusual: unlike normal teeth, the nerves that serve it are inside out. The centre of the tusk is hard, while the outside is covered by softer pulp tissue, thought to be part of a sensory system that detects temperature, pressure changes, chemical gradients and so on. The narwhal also moves in a strange way. It sometimes swims upside down, which is odd because, like all whales and dolphins, its blowhole is on the top of its head. The reason for the inverted swimming seems to be related to the angle of the tusk. It points slightly downwards, an adaptation to swimming under pack ice. However, when the narwhal is travelling close to the sea floor, where it feeds on flatfish down to a depth of 1,500m (4,921ft), there is the risk of the tusk catching on the seabed, which would result in damage not only to the tusk itself, but also to the animal's lower jaw. So, instead of swimming the right way up, it turns upside down.

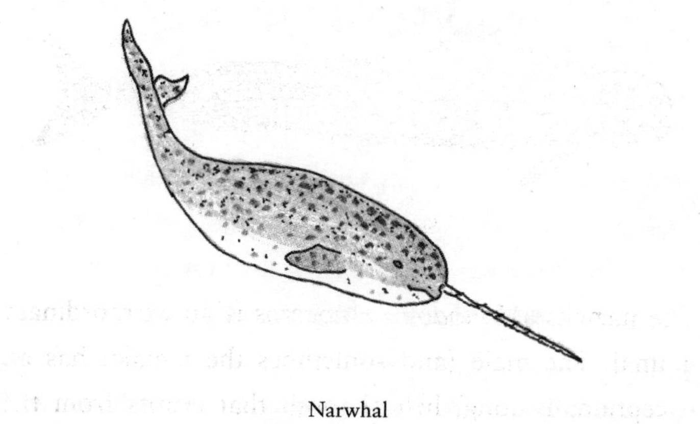

Narwhal

WOLVES OF THE SEA

Orcas or killer whales *Orcinus orca*, which are actually the largest and fastest of the dolphins, are found in all the world's oceans and wherever they occur they look very similar, although this is an illusion. Orcas in different parts of the world have subtle physical differences, such as the size of the white spot behind the eye and the shape of the dorsal fin, and very distinct behavioural differences – so different, in fact, that scientists are considering assigning them to different species. Off the west coast of North America, for example, there are resident pods which patrol specific territories in inshore waters and feed noisily and almost exclusively on fish, especially salmon. Moving in from time to time are transient pods or Bigg's killer whales (named after orca researcher Michael Bigg), which travel through the same inshore areas but feast silently on marine mammals, particularly seals and sea lions. Offshore, there is a third group that specialises in catching schooling fish and sharks. In the Southern Ocean, close to the Antarctic continent, orcas show similarly distinctive feeding patterns: type A orcas hunt down minke whales *Balaenoptera bonaerensis*; type B specialise in seals, but occasionally go for minke whales and humpback whales; type C catch only Antarctic toothfish *Dissostichus mawsoni*; and type D also catch fish, including the Patagonian toothfish *D. eleginoides*.

In at least two places in the world – at Peninsula Valdes in Argentina and the Crozet Islands in the southern Indian Ocean – killer whales pluck seals and sea lions from right off the beach by deliberately stranding themselves. Youngsters practise the technique when about four years old, learning from their mothers. Sometimes things can go wrong but the mother is always there to sort things out. On one occasion at Possession Island in the Crozets a youngster made a lunge for the beach but became well and truly stranded. His mother promptly swam 50m (164ft) out to sea, turned sharply and stormed in towards the beach, creating a large wave that washed the little one back into the water.

Off New Zealand, killer whales specialise in hunting stingrays. They have learned to blow bubbles to drive the rays from their hiding places. They flip them over, which serves to put the rays into a trance, and then tear them apart.

Bull killer whales are mummy's boys. When the time comes to breed, males leave their home pod to mate with females in another pod, but always return to their home pod afterwards, for they are dependent on their mother for their survival, especially males over thirty years old. Females, interestingly, live to ninety or more but stop reproducing at thirty to forty; this is the only animal known to have an extended menopause, apart from pilot whales and humans. It seems females go through the menopause to ensure the survival of their sons, and hence their own gene line, but without the bother of looking after grandsons and granddaughters as well. By investing in her son's well-being, a mother can have more offspring for less effort. She will still have to help bring up her daughters' offspring, the progeny of males from other pods; but the turnover is relatively slow, for dependent calves are not weaned until they are at least two years old, and females breed roughly once every five years and have about five offspring during their lifetime. Her son, on the other hand, will be planting his seed (and her genes) every breeding season, so there is good reason to keep him alive as long as possible. Mothers, therefore, help ageing sons to forage and will defend them in fights. However, it does mean that if a mother of a male older than thirty should die her son inevitably dies soon after. Female offspring are affected less strongly.

Killer whale

One of the most extraordinary relationships between animals and people occurred in Twofold Bay, New South Wales, where a pod of killer whales led by 'Old Tom' drove humpback and fin whales close to shore where the local whaling community at the township of Eden would help to kill them. The dead whale sank initially to the sea floor, where the orcas took their prize – the tongue and lips – leaving the rest of the whale to float back to the surface and be towed to shore by the whalers. This partnership lasted many years, from the 1870s until Old Tom's death in 1930. His skeleton can be seen in Eden Museum.

FLIPPER AND FRIENDS

Bottlenose dolphins *Tursiops truncatus* call each other by name, but only if they are close friends. Each dolphin has its own signature whistle, the equivalent of a name: in fact, it is an abstract name invented by the individual, a phenomenon that until now has been the preserve of only one other animal – us. The dolphin makes this whistling sound when greeting other dolphins. It is telling them, among other things, who it is. Sometimes another dolphin will copy the other's signature whistle, but only if the two of them are close associates, such as a mother and calf or a male alliance, for only animals that have spent time together can share a name. They call if they have been separated for a short while and they want to see each other again. It is as if the other dolphin is addressing an old friend. If two unfamiliar dolphins cross paths the signature whistle is not copied. The research by a consortium of British and American scientists is one more piece of evidence indicating that dolphin communication is closer to that of humans than to any other species.

Two groups of bottlenose dolphins living near Cedar Key in Florida are a bit like lions and chimpanzees in that some members of a hunting group have specialised roles. One dolphin, and it's always the same individual, acts as the 'driver'. It herds shoals of fish in a circle and towards a barrier of half a dozen 'non-driver' dolphins, which line up with their bodies almost touching. The driver in one group has been seen to slap the surface forcibly with its tail, a behaviour that's become known as 'kerplonking', and serves to frighten fish from their hiding places. The driver in the other group drives fish without the tail slaps. Whichever technique is used, the result is the same: the fish attempt to escape by leaping into the air. The driver dolphin surfaces close to the barrier and all the dolphins pick off their quarry with consummate ease, catching the fish in mid-air.

Another group of bottlenose dolphins living in and around the sandbanks and pools of Florida Bay have come up with a slightly different approach to herding mullet. Working in just a metre of water they have a technique known as 'mud-ring feeding'. One female, always the same individual, swims in an anticlockwise circle, flipping up the mud from the seabed as she goes. This acts like a wall, trapping the fish. They try to escape by leaping out of the circle and the dolphins are there ready to catch them.

In South Carolina and Georgia, bottlenose dolphins have evolved yet another feeding technique called 'strand feeding'. The dolphins herd fish towards the shore and create a surge wave that washes the shoal onto the mud bank. The dolphins follow in a mighty whoosh, turning on their sides to grab the stranded fish. Herons have learned that by watching the dolphins' behaviour they too can bag a catch. They simply wait in the right place on the shore to intercept the fish pushed up by the sudden rush. They grab what they can before any surviving fish wriggle back into the water.

In Australia, bottlenose dolphins chase fish into empty conch shells. They pick up the shells and shake them about until the disorientated fish falls out ... right into their open mouths. They also place pieces of sponge on their snouts to protect them from sharp stones when digging for fish beneath the sand. Clever animals, dolphins!

VIOLENT KILLERS

Bottlenose dolphins are killers – not in the predatory sense, but in acts that, in human terms, are tantamount to murder. The victims are harbour porpoises *Phocoena phocoena*, which in some parts of the world, such as off the east coast of Scotland and along the California coast, share the same patch of sea as the dolphins. Until 1995, when the first evidence of porpoise battering in Scotland's Moray Firth hit the headlines, bottlenose dolphins were often depicted as gentle, smiling creatures, but the more they encounter porpoises the more their darker side appears. On the California coast, about fifty fatal attacks of 'porpicide', as it's become known, were reported between 2005 and 2009; but these are probably just a fraction of the actual deaths caused by bottlenose dolphins, which actually seem to 'enjoy' their violence and 'play' with their victims. The scoundrels appear to be testosterone-induced males taking out sexual frustrations on their smaller cousins.

Female sand tiger sharks *Carcharias taurus* mate with several males and may carry the embryos of more than one father. However, the strongest and fastest-growing offspring, probably from the most aggressive and strongest fathers, turn on their womb mates and eat them. Only two well-armed and actively hunting embryos survive to be born, one from each uterus, the result of intrauterine cannibalism.

During the summer of 2007, something unusual happened at the colony of common guillemots *Uria aalge* on the Scottish Isle of May. The normally peaceful community was wracked by mass murder. Two-thirds of the season's chicks were brutally pecked to death, not by the marauding gulls – always a threat – but by the guillemots themselves. A food shortage meant that parents were both at sea foraging at the same time. When single chicks went to be temporarily adopted by a neighbour for food or safety (normal behaviour in the guillemot colony) they were given short thrift and either killed on the spot or tossed over the nearest ledge.

JONAH AND THE WHALE

The story of Jonah and the whale appears in the Old Testament of the Bible and tells the unusual tale of someone who was believed to have lived about 800–780 BCE. He was on the run for not undertaking a task set by God and his ship ran into a storm. When the crew found that he had crossed the Almighty they threw him overboard and the storm subsided. Jonah, meanwhile, was not allowed to drown. As he was sinking, a huge sea creature, described in some Bible translations as a 'whale' and in others as a 'great fish' or 'sea monster', swallowed him up. He was confined inside its belly for three days and three nights, during which time he had the presence of mind to compose a psalm, or so the story goes. When God was assured that he was repentant, the creature spewed up Jonah and he was able to reach the nearby shore, thought to be not far from Iskenderun in southern Turkey. Religious scholars have long debated whether Jonah's story is historical or symbolic; after all, 'whale-swallows-man' stories occur in Greek legends as well as folktales and myths from many parts of the world, and there are even claims that some events really happened. Whales and their propensity to swallow and regurgitate people has been a popular topic for tall tales down the centuries, but let's explore whether a whale or any other giant sea creature really would, or even could, swallow a person, and whether that person would have any chance of survival. A skim through the literature reveals at least one tale that suggests they could. Some devout Christians and creationists often quote it as proof that all events in the Bible are true. The story was reported in such august organs as the *New York Times* and *Princeton Theology Review* and featured in several learned books. It involved a large sperm whale *Physeter macrocephalus*, and the tale goes something like this:

In February 1891, the sailing ship *Star of the East* was a few hundred kilometres to the east of the Falkland Islands when the lookout spotted a sperm whale about 4.8km (3mi) away on the starboard quarter. Longboats were launched in pursuit of the whale. The first boat to close in on the whale speared it with an iron harpoon. It sounded, pulling the boat some 8km (5mi) on a hair-raising excursion that was known by Yankee whalers as a 'Nantucket sleigh ride'. It then turned and headed back to where it was first harpooned. As soon as its back broke the surface, the crew of the second boat let loose another harpoon. Understandably, the whale thrashed about and then struck off on another jaunt for 4.8km (3mi) and dived again. The lines went slack. As the crews were pulling them in, the whale surfaced. It struck one boat with its nose and capsized it, tossing its occupants into the water. The other longboat picked up survivors, but two of the crew were missing, presumed drowned.

An hour or so later, the whale died from its harpoon wounds and the ship's crew brought it alongside and hacked away the valuable blubber. The following morning, they hoisted the whale's stomach onto the deck but were startled to find that there was something moving inside. They cut it open, and there before their eyes was one of the missing sailors. He was doubled up and unconscious, but still alive. They bathed him in seawater and he came round, but for some time he behaved like a raving lunatic. He was taken to the captain's cabin where he eventually recovered and was back to work after three weeks. His skin, though, bore the scars of his curious adventure. Where the whale's gastric juices had started to work, his face and hands were bleached white and his skin wrinkled.

Later, he described how he was lifted up by the whale and slipped down a smooth passage. His hands came into contact with a slime that shrank from his touch. He could breathe but was overcome by the heat and fainted. The next thing he recalled was waking up in the captain's cabin. When the ship returned eventually to England, he was treated for his skin condition at a London hospital.

His name was James Bartley. It is said that he survived to a ripe old age, but his skin was always yellow and like parchment. He became a whaling legend but eventually left the sea and set up a cobbler's shop in Gloucester, where his gravestone bears the epitaph 'James Bartley, 1870–1909, a modern Jonah'. His story is a terrific yarn, but there's a fundamental flaw: little if any of it appears to be true.

The story went into the history books and has been copied from book to book, and nowadays from one website to the next without anyone checking its authenticity. But then Professor Edward B. Davis of Messiah College, Grantham, Pennsylvania, came along. He looked at the evidence and found it wanting. He felt the story of John Bartley, the supposed modern Jonah, should be consigned to maritime mythology, not history. This is what he discovered:

The *Star of the East*, according to *Lloyd's Register of British and Foreign Shipping,* was a real ship, a barque made of iron. It was 183ft long and 734 tons net, and was built in Glasgow by C. Connell and Co. It was owned by Sir Roderick Cameron and registered in London. Its master when commissioned was Captain W. Esson, but at the time of the supposed Bartley story Captain J. B. Killam had taken over. *Lloyd's* recorded that the *Star of the East* left Auckland, New Zealand, on

27 December 1890, bound for New York, where she arrived on 17 April 1891. She might well have been off the Falklands in February of that year, but she was a merchant ship, not a whaler, and John Bartley was not registered as one of the crew. And to confound the story even further, some time later in 1907 the wife of the ship's captain stated quite categorically in a letter published in the *Expository Times* that 'There is not one word of truth to the whale story. I was with my husband all the years he was in the *Star of the East*. There was never a man lost overboard while my husband was in her. The sailor has told a great sea yarn.'

Even more significant is the fact that commercial whaling did not start in the Falklands area of the South Atlantic until 1909 and, according to whaling historians, in those days it is unlikely the stomach of a whale would have been hoisted onto the deck; it would have been discarded. It was only in more recent years when whales were hauled aboard factory ships that scientists examined the bits less interesting to whalers, such as the stomach where they might find pieces of giant squid.

So, if the John Bartley story is fiction, how did it arise in the first place? Davis found that the story first appeared in an edition of the *Yarmouth Mercury* dated 22 August 1891. The *New York Times* picked it up some years later, publishing it on Sunday 22 November 1896 (with a credit to the *Yarmouth Mercury* of South Yarmouth, Massachusetts). The whole joke, however, probably had its origins in a whale story from Great Yarmouth that appeared in the *Yarmouth Independent* in June 1891. It told of a 9m (30ft) long rorqual (baleen whale) that came close to shore. Several boats were chasing it and it hit the pier at Gorleston, just to the south of Great Yarmouth. It ran aground and died. It was then stuffed by a taxidermist and transported around the country on a specially built horse-drawn trailer accompanied by the East Anglian naturalist Arthur Patterson. The Yarmouth lifeboat crew took it to the Earls Court Exhibition, and it was eventually displayed at the London Westminster Aquarium, where it was known as 'The Gorleston Whale'. Newspaper reports at the time mentioned that the whale had inspired several exaggerated stories. Davis suggests that the John Bartley story was simply one of them.

There is, however, a curious postscript to the saga. Captain J. B. Killam was born in Yarmouth – not Yarmouth, England but Yarmouth, Nova Scotia – so maybe there is more to this story than meets the eye.

MORE MEN OVERBOARD

The John Bartley story is not the only one of its kind. There is a tale told by Edgerton Y. Davis, a surgeon with the harp seal fleet out of Newfoundland. He wrote (in 1947) about a young sealer who fell from an ice floe off St John's, Newfoundland, in 1893. An enormous bull sperm whale, which was apparently lost and in Arctic waters out of season, swallowed him. The whale was caught and when its stomach was opened up the man was found inside. He was badly crushed 'in the region of his chest', according to Davis, and partially digested. The gastric mucosa (the lining) in the whale's stomach had encased his body, yet some lice on the sailor's head were still alive. It was, perhaps, another entertaining story, but curiously Davis waited for more than half a century to tell it. Victor Scheffer, the award-winning author of *The Year of the Whale* – an imaginative book on sperm whale biology – is sceptical. He quotes the story but 'ventures to doubt it'.

There is also an even earlier case. It was reported in the *Massachusetts Gazette, and the Boston Post-Boy and Advertiser*, dated Monday 14 October 1771. The report described how a whaling ship arrived at Edgartown carrying Marshal Jenkins, a whaler whose own boat had been bitten in two by a sperm whale. Jenkins was thrown into the water and then caught in the whale's mouth. The whale pulled him below, but later rose to the surface and discarded him. He was taken on board the mother ship, where he was found to be much battered and bruised, but after a couple of weeks he fully recovered.

JONAH, THE WHALE AND REALITY

So, back to the biology – could a sperm whale swallow a person whole? Sperm whales take large prey, such as giant squid and deep-sea sharks. We know this because scientists assigned to whaling factory ships used to find the remains of these creatures inside sperm whale stomachs, and nowadays whales washed up on the shore have been found to have feasted on them. Squid biologist Malcolm R. Clarke, for example, wrote about a 10.5m (34.5ft) giant squid, measured from the end of its body to the tip of its longest tentacles, almost intact in the stomach of a 14m (46ft) long bull sperm whale caught off the Azores. But could such a whale swallow a man?

Roy Pinney tells in his *Animals of the Bible* (1964) of a director of a natural history museum who was often asked if the story of Jonah was true and if a man could be swallowed by a whale. He responded by trying to push his body down the throat of a dead and rather smelly 18.2m (60ft) long sperm whale that had been stranded, and reported that although he could just about squeeze through, a fat man could not have made it. The only conclusions we can glean from that experiment is that

Jonah was of modest build! However, in the *Daily Mail* of 14 December 1928, G. H. Henn, from Birmingham, England, recalls how at the turn of the century the carcass of a large whale was displayed for a week on some wasteland in Navigation Street, outside New Street Station. Henn and eleven other men entered its mouth, passed down its throat and moved about in a chamber that was 'equivalent of ... a fair-sized room'. He said that 'its throat was large enough to serve as a door'. And Frank Bullen records in his *Cruise of the Cachalot* that a whale was once seen to regurgitate 'a massive fragment of cuttlefish as thick as a stout man's body'. So, maybe the jury is out on Jonah's size after all.

Whales under stress often regurgitate the contents of their stomach. Whalers have noticed that a whale in its death throes will bring up its last meal, sometimes vomiting up huge chunks of giant squid and whole sharks. Jonah's whale might also have become stranded. Sperm whales are mid-ocean animals and when close to shore they sometimes get into difficulties. Most probably their echolocation system is switched off when cruising and they are operating on 'automatic pilot'. When they hit a sandbank in shallow water, they are rudely awakened and their sensory systems are overwhelmed. The result is simple – they are beached, and a person trapped in a stranded sperm whale would have the opportunity to walk out.

So, what are the chances that a large whale was passing Jonah's boat in the eastern Mediterranean? Are sperm whales found here? The answer is 'yes', although this species is not common nowadays. The Israel Marine Mammals Research and Assistance Center reports that single individuals and small pods of up to five whales are sometimes seen in the eastern Mediterranean. At 6 a.m. on 7 April 1996, for example, Yoram and Suzette Greenberg, owners of the yacht *Amonte,* observed a larger pod of eight to ten sperm whales about five nautical miles to the east of Rhodes in the Greek islands. They were able to compare the length of the whales with their 14m (46ft) yacht, and estimated two of them to be 15m (49ft) long. And, from time to time, dead or dying sperm whales have been washed ashore on the Israeli coast, although these are thought to be whales that have strayed from their normal haunts, or carcasses that have drifted long distances across the sea.

Whales more frequently seen in this part of the Mediterranean are fin whales *Balaenoptera physalus*, the largest whale species to be found in these waters. Unlike the sperm whale, the fin whale is a filter feeder, one of the whales whose teeth have evolved into comb-like bristles or baleen that enable it to sieve krill and small fish from the seawater. In this group – which includes the blue whale *B. musculus* and humpback whale *Megaptera novaeangliae* – the throat is narrow, far too narrow to swallow an adult human. There are records of large whales choking to death on nothing more than a seabird. In 1829, for instance, a humpback whale that had washed ashore at Berwick in the British Isles had six cormorants in its stomach and another stuck in its throat. According to observer G. M. Allen, it is thought the seventh bird caused the whale to choke.

As for an explanation of the narrowness of the whale's throat we can turn to none other than Rudyard Kipling. He explained 'How the Whale Got His Throat' in one of his *Just So Stories*, published in 1902. This was also a Jonah-like tale, in which the whale had eaten all the fish in the sea and so thought he'd try humans instead. He swallowed a man named Fitch, who was adrift on a raft, but when inside the whale's stomach Fitch jumped about until the whale could stand it no more, and the man walked out. However, before he left he had cut up his raft into a small grating and lodged it in the whale's throat. From that day hence, according to Kipling, 'the whale could neither cough nor swallow, nor could he eat anything except very, very small fish; and that is the reason why whales nowadays never eat men or boys or little girls' ... or Jonah if it comes to that!

JONAH AND THE SHARK

If a whale was not Jonah's assailant, what was? It could be that the monster was a one-off. After all, according to the scriptures God had 'prepared' it to swallow Jonah. Also, the Bible was written in Hebrew, Aramaic and Greek so translators from those languages into English could have opted for 'whale' simply because it was the largest animal known at the time. However, in some translations a much more fearsome creature is offered as Jonah's monster. Photius of Constantinople (c. 810–893) writes of none other than the great white shark, and the Danish missionary Otto Fabricius (1744–1822) adds 'its custom is to swallow down dead and, sometimes also, living men, which it finds in the sea'.

There is no doubt that the great white shark *Carcharodon carcharias* can be a giant among fishes. It is the largest predatory fish in the sea and is the shark that has clocked up the greatest number of recorded attacks on people with the largest number of fatalities (see pages 97 and 113). As we have seen, large specimens have been reported reliably to grow up to 6m (20ft) (see page 98) and there have been unsubstantiated claims for even bigger ones.

Bite marks on whale carcasses off southern Australia suggest there are great whites in that part of the ocean that are up to 8m (26ft) long, although no one can be sure until they find and measure a shark of that length. In 1987 Peter Riseley claimed a 7m (23ft) shark at Kangaroo Island, Australia, but it was not properly measured. Way back in 1894, Captain E. S. Elkington chanced upon a great white that was estimated to be 1.2m (4ft) longer than his 10.5m (35ft) launch. He watched the beast for about half an hour just outside Townsville in Queensland, Australia. An even larger shark was claimed in June 1931, when an 11.3m (37ft) specimen was trapped in a herring weir in New Brunswick, Canada. But South African writer Lawrence Green trumps them all. In his book *South African Beachcomber* he mentions a white shark that was reputed to be 13m (43ft) long. It was found at False Bay, near Cape Town, at the end of the nineteenth century, not far from Fish Hoek Beach where an exceptionally large great white shark, described by eyewitnesses as 'like dinosaur huge', killed and then swallowed a holidaymaker as recently as January 2010. His swimming goggles and a patch of blood were all that remained.

The largest reliably measured great white of modern times, however, could be a female caught off Filfla, a small flat-topped island to the south of Malta. It was April 1987 … and the story takes us firmly back to Jonah and the Mediterranean. Photographs of her were taken on the quayside at Marsaxlokk, and based on those pictures Alessandro de Maddalena of the Italian Great White Shark Data Bank and his colleagues estimate that the shark was between 6.7 and 6.8m (22ft) long. She had a blue shark, a large swordfish and a dolphin in her stomach. Later the same year, two white sharks became entangled in nets off the Egadi Islands on the west coast of Sicily; one was a 5m (16ft) long female.

All three of these enormous sharks were found in a stretch of the Mediterranean known as the Sicilian Channel, between Sicily and Tunisia, a known white shark hot spot. Since the late nineteenth century there have been hundreds of encounters; in fact, about 40 per cent of great white shark sightings in the Mediterranean and the north-east Atlantic have been here. Many have been caught, most of them big fish. According to de Maddalena, five known specimens have measured over 5.9m (19ft) and at least two have been more than 6m (20ft) long. However, in the summer of 2009, a newly born 1.5m (5ft) long great white was found in the net of a trawler out of Lampedusa, a small island between Malta and Tunisia. It was an indication that the large, mature females are possibly in the Sicilian Channel for one thing – to give birth. The channel could be a pupping site and nursery for white sharks not only in the Mediterranean but possibly the north-east Atlantic as well. At the beginning of 2012, a 4m (13ft) long female was brought into the Moroccan port of Kenitra on the Atlantic coast ... was she on her way to the Med? We'll never know, but sharks heading into the Mediterranean, which could well have slipped in from the Atlantic, are seen on both the north and south sides of the sea.

On the northern shore of the Mediterranean, a monster white shark was caught off Sète, to the south-west of Montpellier in the south of France. The date was 13 October 1956. The shark was 5.83m (19.13ft) long, and a cast of her can be seen in the Museum of Zoology in Lausanne, Switzerland. White sharks in this north-western part of the Mediterranean are surprisingly common. According to shark enthusiast David Diley, several white sharks have been spotted off Barcelona and Valencia in recent years, and cruise ship crews, crossing from Ibiza to Nice, report seeing a couple of great whites at the surface every year. Some of them are undoubtedly big. Diley heard from a fisherman, who had once caught a giant great white, that in September 2010 he found a tuna with a very large and characteristic white shark bite floating in the sea off Mallorca.

It is thought that these sharks on the northern shores of the Mediterranean are heading not to the Sicilian Channel, but swimming generally towards the eastern Mediterranean, hugging the coasts of France and Italy. You can track the possible route by the records, both ancient and modern, of white sharks in this part of the sea.

In the summer of 2012, for example, a great white was filmed about 300m (984ft) from the shore in the Gulf of Saint-Tropez. On 16 March 1954, a female estimated to be 7m (23ft) long, but whose actual dimensions are in question, was taken off Camogli, to the east of Genoa. In 1886, a truly monstrous white shark, estimated to be between 8m (26ft) and 10m (33ft) was caught near Piombino, opposite the island of Elba, on the border of the Ligurian and Tyrrhenian seas. And, on 18 September 1979, a 6.2m (20ft) long white shark was taken near Gallipoli on the 'heel' of Italy, which is close to the entrance of the Adriatic Sea. This is significant, for the Adriatic is another white shark hot spot.

The giant sharks are here right enough, although their actual size might be questionable. In 1891, for instance, an enormous great white was caught during naval manoeuvres in the Adriatic Sea. It was estimated to be a staggering 10.06m (33ft) long and weighed four tons. Even if the length turns out to be an exaggeration, it was probably still a whopper. And these monsters appear on both sides of the sea. Tracking the occurrence of the sharks on the Italian side only, most sightings have been in the northern sector of the Adriatic.

On 27 October 1999, skipper Elvio Mazzafugo and his crew on the 10m (33ft) long fishing boat *Coca Cola* encountered a female great white, estimated to be 5–6m

(16–20ft) long, off Giulianova. They were fishing for bluefin tuna *Thunnus thynnus* in water about 250m (820ft) deep. The crew had just hooked a fish, which was tied alongside the boat, when the shark appeared. It savaged the tuna and attacked the boat. The encounter followed the sighting of a large shark, probably a great white, in the Gulf of Trieste in the north-east Adriatic a few days previously.

In 1839, a giant turned up at Civitanova Marche. Its length, based on the size of its largest vertebrae, was thought to be around 6.02m (19.75ft).

Further to the north, on 27 August 1998, a large female great white, estimated from video footage to be about 5.5m (18ft) long, approached a fishing boat about 35km (22mi) off the coast at Senigallia, between Ancona and Rimini. A sports angler on board was fishing for sharks in about 70m (230ft) of water, and had caught a thresher shark *Alopias vulpinus*, which he had tied up alongside. Some time later a very large great white appeared. It circled the boat and grabbed the burlap sack hanging from the stern and then took a half-metre-wide chunk out of the thresher. It also managed to take some glass fibre and teak pieces from the boat's hull. It was, according to Ian Fergusson – a British shark enthusiast with a special interest in great white sharks in the Mediterranean and north-eastern Atlantic – one of the largest great whites ever captured on video, and it was a female. From the pictures, he estimated that she weighed about 2,000kg (4,409lb) and her largest teeth were 5cm (2in.) long from root to tip.

A little further again to the north, a large great white nicknamed 'Willie', but in reality a female, turned up in several places in the Adriatic during the summer of 1989. She was estimated to be over 5m (16ft) long. In September that year she surprised sports anglers off Pesaro, to the south-east of Rimini on the Italian coast, where she circled boats and broke fishing lines.

These are just a few of the many sightings recorded on both sides of the Adriatic, and there have been attacks on people too, some fatal; but why are these large and aggressive females here at all? Could the Adriatic be another shark nursery in the Mediterranean region? The answer, again, is probably yes.

In December 1991, commercial fishermen caught a 2.1m (7ft) youngster off Ancona, and in March 1992 a 2.3m (7.5ft) long specimen was caught in nets off Termoli – both towns on Italy's Adriatic coast. These small white sharks could not have travelled far; they were still in their nursery areas. In addition, in 2008 two newborn great whites were caught off Altinoluk on the Aegean coast of western Turkey. The smaller shark was just 1.25m (4ft) long, the smallest found to date in the Mediterranean.

But, how about further east? What about Jonah's sector of the Med? Could large female great white sharks (the kind that could swallow somebody whole) be in the far east of the Mediterranean ready to give birth? The evidence is scanty, but again the answer could be yes. The scarcity of great white shark sightings in the Levantine Basin in the south-eastern Mediterranean, suggests Ian Fergusson, is due in part to higher temperatures and salinities. Nevertheless, from time to time there are sightings and captures.

In 1934, a gravid female was caught at Agami Beach, near Alexandria, indicating perhaps that nursery sites could also be or could have been close to the Levant. It took three boats' worth of fishermen to haul her ashore. She was estimated to be about 4.3m (14ft) long and weighed 2,540kg (5,600lb). When she was cut open, nine 60cm (2ft) long embryos were found inside. This puts the great white shark firmly in the right area for Jonah's abduction. Also, there have been sightings of great whites off Cyprus, Lebanon and Israel, as well as shark attacks by species unknown off Greece (six) and Lebanon (three).

People, however, are not the primary targets of great white sharks. Great whites in the Mediterranean hunt or scavenge a variety of foods – bottlenose, striped and common dolphins, tuna, bonito, swordfish, blue and shortfin mako sharks, rays, sea turtles, dead whale and dolphin carcasses and dead domestic mammals that have drowned or been washed into the sea. But should one accidentally or deliberately swallow a person, what are the chances of them being vomited up again?

Sharks *do* regurgitate food items, such as human arms and legs, which they bring up when under stress – but an entire human? To date there are no records of them doing so, although there was a historical account of an incident that came close. In 1758, according to E. B. Pussey in his book *The Minor Prophets* (1886), a sailor fell overboard from a frigate that was sailing in the Mediterranean in stormy weather. A monster white shark, which was known as a 'sea dog' in those days, took the unfortunate seaman into his wide throat and swallowed him whole. His fellow crewmen leaped into the ship's sloop and tried to help their comrade. The captain of the ship ordered a small cannon to be fired at the giant fish, which must have been thrashing around at the surface. It was struck by the cannonball and promptly vomited up the sailor, who was still alive and 'little injured' ... or so the story goes – nobody has checked this one either. The sloop picked him up and the shark was harpooned. It was estimated to be about 6m (20ft) long and weighed 1,780kg (3,924lb), quite a size for a great white. It was dried and presented by the captain to the sailor. He then went around Europe, it is said, exhibiting both the shark and himself. It moved one naturalist to write: 'It is probable that this was the fish of Jonah.'

More usually, the bodies found inside sharks are dead. On 2 October 1954, for example, a 0.9 tonne (2,000lb) great white caught off Nagasaki, Japan, was found to have a thirteen-year-old boy in its stomach, and it is not unusual for these giants to contain even bigger corpses.

There are a couple of historical accounts of giant sharks taking whole human prey. In 1776, Welsh naturalist Thomas Pennant described how a great white shark was caught with 'a whole human corpse ... in the stomach'. In 1909, three partially clothed human cadavers – an adult male, adult female and a young girl – were found in the stomach of a 4.5m (15ft) female great white shark that was caught in fishing nets off Cape Santa Croce, a peninsula between Syracuse and Catania in Sicily. They were thought to be earthquake victims that the shark had scavenged after a tidal wave, triggered by the tragic Messina earthquake of 1908, had washed them out to sea. There's even an account of an entire horse being found in a Mediterranean white shark's stomach.

Ian Fergusson feels it is not stretching the imagination too much to think of a great white shark being caught, hauled ashore, eviscerated and a whole human corpse falling out onto the fish market floor; and that would certainly have caused a stir. It is not hard to imagine, then, a local folk-tale emerging that became ever more surreal every time it was told. Could this have been the origin of Jonah's story?

Ancient Greek writers such as Aristotle and Herodotus knew of the great white shark. They referred to it as the 'lamia' monster and even today Greek and French fishermen know the shark as 'lamie'. Herodotus tells how thousands of soldiers and seamen were 'seized and devoured by monsters' when the Persian battle fleet was destroyed in a storm near Mount Athos in the northern Aegean Sea in 492 BCE. And one of the earliest shark attacks on pictorial record was discovered in the Mediterranean. It was found on the sides of a vase dated about 725 BCE. Fragments of the vase, excavated at Lacco Ameno on the Italian island of Ischia, depict a shipwrecked sailor being devoured by a fish, probably a shark.

Pliny the Elder (CE 23–79) mentions sponge divers in the Mediterranean caught up in 'frantic combats with sea-dogs which attack the back, the heels, and all the pale parts of the body'. And the Greek poet Leonidas of Tarentum describes how a sailor was bitten in two by a sea monster. The man's companions interred what remained after the attack; thus he was buried, according to the poet, 'both on land and in the sea'. Records show that in medieval times, great whites were caught occasionally between Sète, where the Lausanne shark was located, and Nice.

However, the notion that a shark was Jonah's nemesis, rather than a whale, has been mentioned many times.

In 1566, French naturalist Guillaume Rondelet, from Montpelier, put forward the notion that a great white shark and not a whale swallowed Jonah. He risked the condemnation of the Vatican by suggesting that a whale's throat was too narrow, whereas the great white shark had the capability of swallowing Jonah whole. His proposition was based on reports of a shark containing 'two tunny and a fully clothed sailor'. The tunny were probably the enormous bluefin tuna up to 3m (10ft) long. In the middle 1700s, the Swedish naturalist Carl Linnaeus, the man responsible for today's system of animal and plant classification, put forward a similar proposition. Rondelet, though, went one better. He wrote about whole men, complete with armour, found inside large great white sharks. They were caught close to Nice and Marseilles, although today maybe his tales are a touch too much for *us* to swallow.

In fact, wearing armour might be the only way to survive passage in and out of a white shark's mouth, for squirming past the shark's extremely sharp, triangular teeth unharmed would be a greater miracle than being swallowed and spewed up by a whale. Up to twenty-eight full-size teeth in the top jaw and another twenty-five in the lower are backed up by rows of smaller, slowly growing teeth that move forward to replace the front set when they drop out. Although whole seals, porpoises and other sharks have been found in white shark stomachs, the manner in which the shark feeds would mean the bodies would be severely lacerated.

On contact with the prey, the lower jaw moves forwards and upwards, impaling the prey on the slightly narrower bottom teeth. Then the upper jaw drops as the mouth closes. If the prey is very large, such as a whale carcass, the top teeth are used to carve through blubber or flesh, the shark shaking its head from side to side to help the slicing. If the prey can be swallowed whole, the movement between the top and bottom jaws gradually draws the victim into the mouth, the shark's teeth keeping it from escaping. Each tooth is serrated, like a saw, and together an entire mouthful can slice through skin, flesh, bone and even turtle shells. There was an unusual shark attack tale that illustrates their slicing potential.

The incident took place near Loch Lomond ... yes, the freshwater loch north of Glasgow in Scotland, and the shark was in a pub; at least, its jaws were. A female customer was larking about and put her hand into the open maw but caught her fingers on the teeth. She was badly lacerated and had to seek medical attention. It was one of the more unusual entries in the International Shark Attack File!

JONAH AND A MONSTER SHARK

Megalodon

The chance of Jonah entering or leaving unscathed through the jaws of a great white shark would have been slim, but there was once a creature that could have swallowed half a dozen Jonahs in a single gulp and they would not have touched the sides. It was a giant shark, the largest that ever graced the oceans – the ultimate shark. It is known popularly as 'megalodon', and scientifically as *Carcharodon* (*Carcharocles*) *megalodon*. It is only known from its fossil remains, in particular its teeth, and they are truly enormous. A single front tooth can be more than 15cm (6in.) tall – yes, more than the width of a hand – with serrated edges like the teeth of the great white shark. Depending which theory you buy into, *C. megalodon* was related either to the modern great white or to mako sharks *Isurus*. It had its heyday in the Miocene and is thought to have died out about one-and-a-half million years ago, but there are a couple of stories that suggest (albeit rather contentiously) it could have lived into more recent times. What is to follow may seem fanciful, but again, let's have some fun and speculate.

The first event took place sometime between 1873 and 1876 when the British research ship HMS *Challenger* was sailing the oceans on the world's first global oceanographic expedition. The crew dredged up fossil megalodon teeth from the floor of the southern part of the South Pacific Ocean at station 281 from a depth of 4,300m (14,108ft). When the scientists examined them they found that the teeth were what they described as 'fresh' – that is, fresh geologically speaking. Two teeth were examined. One was thought to be approximately 24,000 years old and the other 11,000. This meant that the owner of the younger tooth was swimming the oceans at the same time people were crossing the Bering land bridge from Asia into North America – in other words, the shark would have been contemporaneous with humans.

And *Challenger* was not the only research vessel to dredge up large and apparently recently shed sharks' teeth. In the Pacific Ocean, the United States exploring steamer USS *Albatross* dredged up teeth that would have matched a shark about 24m (79ft) long. However, since these dramatic revelations were made the *Challenger* teeth have been re-examined and the way in which teeth fossilised has been revisited. The feeling now is that they are much older than was first thought. The bottom of the sea, however, is not the only place to find them. Fossil megalodon teeth are frequently uncovered around the islands of Malta and Sicily, where they were known as petrified 'dragon' or 'serpent' teeth because it was thought they had supernatural properties. However, the paranormal aside, the huge number of fossilised teeth to be found on these islands indicates that the megalodon was present in large numbers in ancient times in the western arm of the Tethys Sea, precursor to the Mediterranean – Jonah's home ground.

The second story is a little more chilling. It appears in D. G. Stead's *Sharks and Rays of Australian Seas* and in the *Barrier Miner* newspaper of Monday 4 February 1918. The location was Nelson Bay, 150km (93mi) to the north of Sydney, New South Wales. It was (and still is) an area that great white sharks frequent, but the shark that the local fishermen encountered was of a quite different order.

The fishermen, many of Greek descent, were fishing for spiny lobsters, known locally as crayfish, and they caught them, up to three dozen at a time, in lobster traps about a metre (3.3ft) across that were placed in deep water. Stead met the fishermen on Nelson Bay Wharf as they were packing them away to send to the market in Sydney, and he asked casually if they had lost pots in the recent blow from the south-east. At this, they became very agitated and, ignoring the storm, they described 'one, big shark', the likes of which the fishermen had never seen before. It took 'pots, mooring lines and all', according to eyewitnesses, and when its size was mentioned, the fishermen estimated that it was as long as the wharf they and Stead were standing on. It was 35m (115ft) long! The water 'boiled over a large space' when the shark swam past, they said, and it had taken such a large bite out of one of the lobster boats that it had to be beached for repairs.

Stead was sure the fishermen were exaggerating, but there was no doubt in his mind that they had encountered a shark of unusual dimensions and they were quite adamant about what they had seen. The fishermen were familiar with whales – they were on a humpback whale migration route – and they often encountered several species of large sharks, including great whites, but they were adamant that their monster was most definitely a staggeringly big shark. Its head, according to one, was 'the size of the roof of the wharf shed at Nelson Bay', but the thing that struck Stead was their description of its colour. To a man, they agreed it was a 'ghostly whitish' colour. So, if these Australian fishermen are to be believed, and there's no reason to consider that they were lying, megalodon or a species of shark of colossal proportions could certainly have been patrolling the ocean in Biblical times.

Interestingly, a local myth reflects such a creature not that far away from the Nelson Bay monster. In the Torres Strait, between Papua New Guinea and Queensland's Cape York Peninsula, there is the legend of Mutuk, who was delivered from the belly of a shark that had swallowed him (sounds familiar?). Although an accepted legend, Mutuk's story is given some degree of credibility by the real events that surrounded Papua pearl diver Iona Asai. In 1937 he was diving in just 3.7m (12ft) of water when a tiger shark *Galeocerdo cuvier* swam rapidly towards him. In an instant, Iona's head was in the shark's mouth, but he survived. He tells what happened:

> When I turned I saw the shark six feet away from me. He opened his mouth. Already I have no chance of escape from him. Then he came and bite me on the head. He felt it was too strong so he swallowed my head and put his teeth around my neck. Then he bite me.

Iona escaped death by feeling for the shark's eyes and pushing hard until it let go. He made for the mother boat, was pulled aboard and then he fainted. His hospital record is there for all to see – he needed 200 stitches to sew up the two rows of tooth marks in his neck, and three weeks after he left hospital an abscess on his neck produced a tiger shark's tooth. His name, Iona, was a direct native translation of 'Jonah', one of Christian Papuans' Bible heroes.

KRAKEN UNCOVERED

The King's Mirror or *Konungs skuggsjá* is an account, written in CE 1250, of a series of conversations between Norwegian King Håkon Håkonsson and his son Magnus Lagabøte (see page 255). It was meant to be an educational dialogue to equip the future king for office and in a section describing the wildlife of the seas around Iceland enthusiastic mention is made first of many types of whales and dolphins. The king becomes more cautious, however, when he speaks of a 'fish not yet mentioned that it is scarcely advisable to speak about on account of its size'. He goes on to say that even fishermen rarely see it, not one has been caught or found dead, and that only two probably exist and they do not breed. He calls it the *hafgufa* or kraken. He says that it is 'more like an island than a fish', and goes on to describe how it feeds: it 'gives forth a violent belch, which brings up so much food'. The monster's open mouth is said to be as wide as a sound or fjord.

Now, is that thirteenth-century description beginning to sound familiar? ... Think 'volcano'. Could the word 'kraken' have been assigned originally to a submarine volcano that has pushed up suddenly and explosively through the sea's surface, and that the kraken's identity has changed down the eons as increasingly more information about another mysterious sea creature came to light?

A volcanic eruption in the sea is a natural event not uncommon off Iceland – witness the birth of Surtsey in 1963 and Jólnir in 1966 – and would certainly have caused considerable disturbance in the water, much like a monster thrashing about. In the sixteenth century the last Roman Catholic Bishop of Uppsala in Sweden, Olaus Magnus (1490–1557), told of people lighting a fire on the creature's back after which it sank beneath their feet, and in 1752 the Bishop of Bergen Erik Ludvigsen Pontoppidan (1698–1764) described the kraken as a floating island. They are both vivid, if a little fanciful, descriptions of a volcanic island in the making.

These two chroniclers, however, went on to describe what they did not know then, but we know now to be something entirely different – the giant squid. In a 1658 translation of Olaus Magnus, he described 'their forms are horrible ... all set with prickles ... sharp horns round about, like a tree rooted up by the roots'. And Pontoppidan describes 'horns which ... stand up as high and as large as the masts of middle-siz'd vessels'. So, in their accounts could we be seeing the transition between volcanic eruption and giant squid? A kraken had stranded on a Norwegian beach in 1680 and some fishermen had a frightening encounter with a kraken, two events mentioned by Pontoppidan which move ever closer to a biological explanation rather than a geological one, but it was not until 1847 that Danish zoologist Professor Johannes Japetus Smith Steenstrup (1813–97) presented a paper to Scandinavian naturalists that the biological version of the kraken scenario was scientifically recognised as the giant squid.

THE WORLD'S BIGGEST INVERTEBRATES

The giant squid *Architeuthis* is one of the world's largest invertebrates (if not *the* largest). The longest known to date was 18m (59ft) long, measured from the tip of its two long tentacles to the tip of its body. This enormous creature, however, even though it has appeared in many novels and sea monster films, had never been seen in its natural environment. Most specimens have been dead or dying bodies washed up on the shore or caught in commercial fishing nets, but in 2004 all that changed when the remote cameras of a Japanese research team from the National Science Museum photographed a giant squid, estimated to be more than 8m (26ft) long, in its normal habitat in the deep ocean.

It was filmed at a depth of 900m (2,953ft) over a steep continental slope off Ogasawara Island. It attacked a bait of Japanese common squid *Todarodes pacificus* by wrapping around the club-like ends of its two long, gangly tentacles, revealing an unforeseen ability. The giant squid is unique in being able to zip together its two long, extendable tentacles, fitting small suckers along their length into corresponding lugs. The result is a single shaft with the two club-like ends forming a giant claw-like structure at its end.

Giant squid

Unfortunately, this particular squid became snagged on the hooked jig but it showed that it was immensely powerful, more so than anybody had ever imagined. For several hours it tried to break free, carrying the entire rig up to a depth of 600m (1,969ft), before sinking back down to 1,000m (3,281ft). It approached the line repeatedly, spreading its arms widely or enveloping the bait, confirming that this species is an active predator and not, as some scientists have suggested, a neutrally buoyant and sluggish creature. The pictures also showed that, when pulling its catch towards its mouth, the squid wraps its long tentacles around it, much like a python throws coils around its prey.

Eventually a tentacle broke free and was hauled up to the surface, but the event had shown how the giant squid behaves, at least for a short time. It had been hunting at depth during the day but, even so, it was below the deepest penetration of light from the surface. It is no coincidence that this is just the place where sperm whales are hunting too ... for giant squid.

Exceptional lengths are known for the giant squid but an even more robust creature eclipses this enormous invertebrate by weight. It is the 14m (46ft) long colossal squid *Mesonychoteuthis hamiltoni*. While the giant squid is slender with relatively smooth suckers on its arms and two long tentacles, the colossal squid is bulky, wider and has suckers rimmed with teeth. In addition each sucker has a sharp three-pointed or a swivelling hook at its centre. It also has the largest eyes of any known animal – the size of footballs – and the largest known beak, which can slice through flesh like a knife through butter. But, even monsters like these are not immune to predation. Three-quarters of the food of bull sperm whales is colossal squid, and scars around the head and body of the whales, thought to be caused by the squid's hooked tentacles, is testament to the gargantuan battles that take place 2.2km (7,218ft) deep in the Southern Ocean.

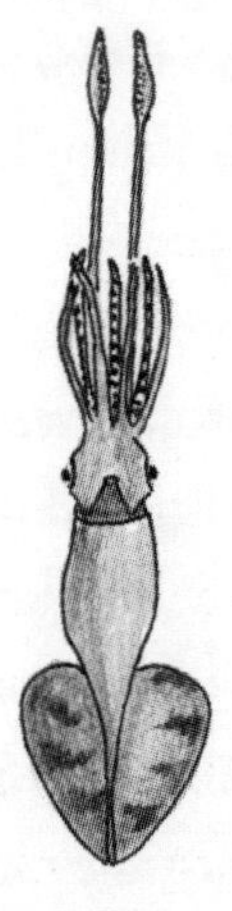

Colossal squid

Humboldt or 'jumbo' squid *Dosidicus gigas* are the gangsters of the cephalopod world. They can jet around at up to 72km/h (45mph), which puts them on a par with the ocean's fastest fish, hunt in large, synchronised groups, 'talk' to each other by changes in their body colour, flashing red and white many times a second like strobe-lights, and can be aggressive to people. The body can grow to 2m (6.6ft) long but are more usually about 1.2m (4ft). They have sucker cups on their tentacles lined with teeth and a razor-sharp beak. They are not man-eaters, however, but catch market squid and fish, such as lanternfish, which they follow during their daily vertical migration down to depths by day and to the surface at night. Baby squid are no bigger than a grain of rice so they must consume huge quantities of food to reach adult size within their two-year lifespan. Many don't make it. Apart from predation by sperm whales, Risso's dolphins, tuna, and mako and thresher sharks, big jumbo squid might also cannibalise smaller ones, their own species making up 15–20 per cent of their diet, so youngsters tend to keep their distance from the adults. They used to be confined to the Pacific coasts of South and Central America, but in recent years they are being found increasingly further north along the west coast of the USA, even as far as Alaska.

ALL HEAD AND ARMS

Octopuses, as we all know, have eight arms, but a study of common octopuses *Octopus vulgaris* at Sea Life Centres in Europe found that when moving across the sea floor they inevitably use the rear two arms as 'legs', leaving the other six arms to search for and grab food. They also use these two rearmost limbs to push off from the seabed and then use the other arms, like fins, to swim slowly.

The blanket octopus *Tremoctopus violaceus* is a pelagic octopus living in tropical and sub-tropical seas. The female, which can be up to 2m (6.6ft) long, is considerably larger than the walnut-sized male – the greatest sexual size difference of any known large animal. At breeding time he has a special arm that stores sperm and, like a scene from the horror movie *The Hand*, this detaches itself and crawls into the female's mantle to fertilise her 100,000 eggs. This species also has an unusual way of defending itself. It is immune to the stings of the Portuguese man o' war (see page 123), so the tiny males and immature females rip off the stinging tentacles and brandish them for their own defence. When harassed the mature female does not hide behind a cloud of ink, like most other octopuses; instead she billows out her arms, which are joined by translucent webbing, to give the impression she is considerably bigger than she really is.

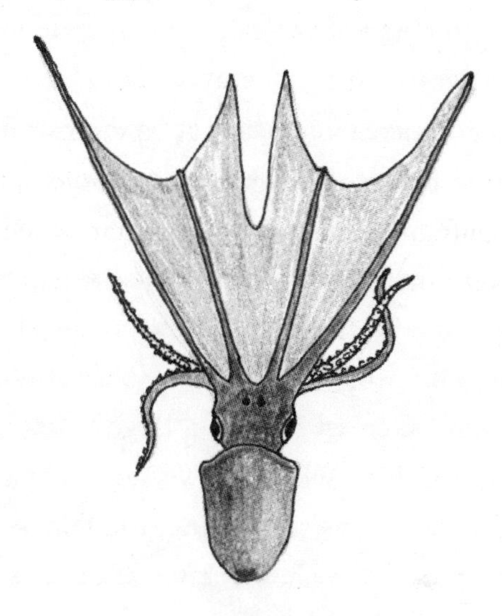

Blanket octopus

When reaching for food, female common octopuses stretch their arms further than males, and smaller octopuses stretch further than larger ones. While smaller, younger octopuses have a greater need of nutrients to grow and are probably more agile than older ones, the researchers at the Scuola Superiore Sant'Anna University in Pisa, Italy, were at a loss to explain the sex differences. All animals tested could stretch their arms to about twice their normal length when at rest, but females went that extra bit further.

The big blue octopus *Octopus cyanea* is not very big and generally not blue. A mature adult's body is at least 16cm (6in.) long and its arms another 80cm (32in.), and like all bottom-dwelling octopuses it can change its skin colour to match its background. One individual living in Micronesia was seen to change its skin patterns 1,000 times while foraging over different types of seabed during a seven-hour period. For the male of the species, this ability to change colour rapidly is a useful skill to have for mating in this species can have unexpected consequences. On one occasion researchers from the Marine Biological Laboratory at Woods Hole, Massachusetts, saw a small male crawl over to a large female and mate with her twelve times while she was foraging away from her den. After that twelfth mating, the female appeared to be hungry and she chased after another small male who barely got away by jetting upwards and producing a cloud of ink. Some time later the first male returned and mated with her a thirteenth time, but for him thirteen was undoubtedly his unlucky number. The female suddenly grabbed him, smothered him, yanked him into her den, and slowly ate him over the next couple of days. The following day another male mated with her, the difference being that he hid well camouflaged behind a rock and only stretched out his arm to deposit sperm in her mantle, so he survived. Watching others, the researchers found that the males preferred to mate in the open, where there was a chance to get away. They also noticed that suitors tended to approach females that had recently fed, thus reducing the risk of being eaten themselves.

Male greater blue-ringed octopuses *Hapalochlaena lunulata* try to mate with any blue-ringed octopus they meet, regardless of sex. The male immediately thrusts his modified third right arm, which delivers packets of sperm, into the mantle cavity of his companion. If it's a male, he withdraws within thirty seconds and the two go their separate ways without fighting. If it's a female they remain together for a couple of hours, after which the female throws out the male forcibly.

A large female football octopus *Ocythoe tuberculata*, caught in 2003 in the Aegean Sea, produced the most eggs of any known octopus – around one million – and it is the only octopus or squid species to give birth. The mother retains developing eggs and the young are not released until they are ready to break out. The species lives in the surface waters of the open sea and maintains its position in the water column with a true swim bladder, like many fish. It is the only octopus known to do so. In addition to its siphon, it also has two additional nozzles high up on each side, which enables the football octopus to control swimming better than other species of octopus. It has four long arms and four short ones, and the 10cm (4in.) dwarf males are considerably smaller than the metre-long females. Young females and males are sometimes found hiding inside salps.

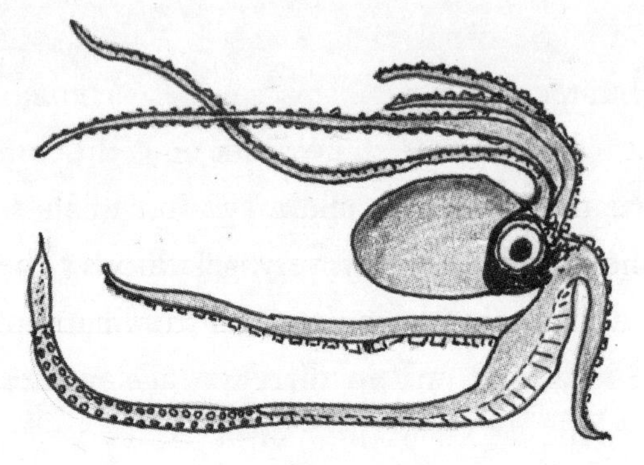

Football octopus

The world's largest octopus is generally considered to be the giant Pacific octopus *Enteroctopus dofleini*, which can be found on rocky shores around the North Pacific. Adults on average weigh about 15kg (33lb) and have an arm span (with the arms laid out and measured from arm tip on one side to the arm tip on the other side) of about 4.3m (14ft), the arms accounting for a little over a quarter of the total length, but much larger ones have been claimed. An individual captured off the coast of British Columbia in 1967 was said to have a radial span of about 7.5m (25ft) and weighed about 70kg (154lb). Measured from the top of the head/mantle to the tip of the arms, record breakers of this size would be about 4m (13ft) long.

A close rival is the seven-armed octopus *Haliphron atlanticus*, a deep-sea species that lives 1,000m (3,281ft) down in the Atlantic and Pacific oceans. Despite its common name it has eight arms, like most other octopuses, but the male tucks his eighth arm into a pouch in front of his right eye. He uses it as a sexual organ when the time comes to mate. The rest of the arms are quite short and the body very gelatinous. The largest known was caught incomplete near the Chatham Islands about 680km (423mi) to the east of New Zealand. It was a female and if she had been intact it is estimated she would have been about 4m (13ft) from the top of the mantle to the tip of her longest arms and would probably have weighed about 75kg (165lb), which puts her on a par with the giant Pacific octopus. Few specimens have been collected, and little is known about the life of this species in the darkness of the deep sea.

SIZE IS EVERYTHING

The largest penis in the natural world is – surprise, surprise – that of the blue whale *Balaenoptera musculus*, yet the statistics are far from certain as little is known about blue whale reproduction and nobody has had the temerity to carry out measurements when the creature is aroused and pursuing a female, so dimensions must be speculative. The figure generally bandied about for a mature male is a length of 2.4m (8ft) and a girth of 30cm (12in.), with an ejaculation load of … wait for it … 20 litres (35 pints), based on the size of a mature whale's testes. The penis is normally held inside the body and pushed out during intercourse.

Size for size, the acorn barnacle, a crustacean that has adopted a sessile lifestyle on rocks, has the longest penis of any known animal. It can be eight times its body size and is used to inseminate neighbouring barnacles that might be some distance away. However, penis size and shape are linked directly to the turbulence of the exposed shoreline. In rough waters the penis is shorter and stouter than that in barnacles in sheltered places, but barnacles transplanted from one area to another can change penis shape to deal with local conditions. Many species of acorn barnacle are hermaphrodites – they start out as females and then change into males. During the July breeding season the female barnacles are receptive for just over an hour, and copulate with up to eleven males up to 582 times. Within three hours, the individual then changes sex and copulates as a male, each insertion lasting no longer than 2.4 seconds. The undue haste appears to prevent the exposed penis from being damaged by wave action or from predation by crabs.

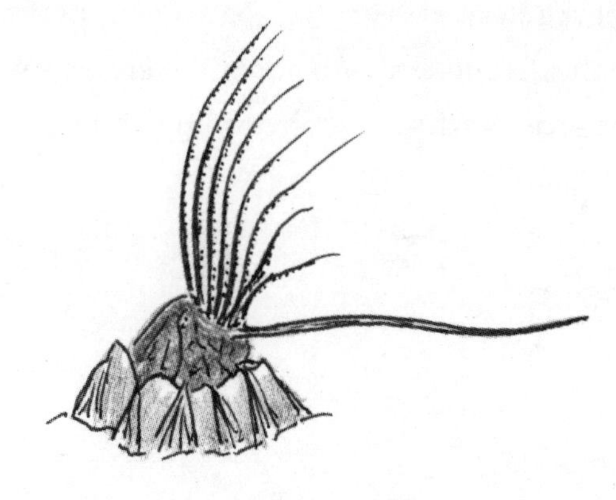

Acorn barnacle

After mating and spawning many species of deep-sea squid simply disintegrate. Their muscles become flabby, their arms and tentacles drop off and they quickly die. Researchers at the University of Otago in New Zealand took a look at what is going on. They studied the deep-sea greater hooked squid *Onykia (Moroteuthis) ingens*, which has some interesting vital statistics. The smaller male has huge testes and a penis that when erect is the length of the arms, head and mantle combined, making it the longest penis relative to body size of any mobile animal. The much larger female has such big ovaries they squash all the organs in the body including the gut so she can't feed. After laying her eggs the gut is unrestricted; however, the female does not feed again for she cannot. The egg-production process activates a gland near the eye that stops the squid from synthesising proteins and her body breaks up and she dies not long after giving birth. It is thought that a single very big spawning is preferable to many smaller events because predators are overwhelmed with food so at least some baby squid survive.

CLEVER CRUSTACEANS

The animals with the world's strongest muscles are stone or rock crabs, genus *Cancer*, which includes Europe's edible crab *C. pagurus* and California's brown rock crab *Romaleon (Cancer) antennarium*. Their claw-closing muscle fibres are made from longer than normal overlapping filaments (sarcomeres) that, when contracted, exert six times more force than even the powerful flight muscles of birds. They use their record-breaking claws to crack hard-shelled prey, which might include barnacles, clams and smaller crabs.

Throughout their lives, spiny lobsters *Panulirus* are great travellers. Right from the word go, the larvae float about in the ocean currents as part of the zooplankton, but they do not journey alone: they hitch a ride with jellyfish, whose muscular bell provides propulsion, and if they should slip off their legs have feathery extensions that help them swim unaided. Once a larva settles on the seabed and grows into a full-size lobster its travels are not over. Each autumn on the east coast of North America, spiny lobsters march across the seabed in long lines, each lobster maintaining contact with one in front using its long antennae. They might cover 80km (50mi), moving over the sandy sea floor at 3km (1.9mi) a day. It's not clear why they make these annual journeys, but there is speculation that they are avoiding the rough winter storms that pound the rocky shores on which they spend the summer. It is thought they use the Earth's magnetic fields to find their way, for the lines of lobsters tend to follow the same compass bearing.

Spiny lobsters

Should a small male American lobster *Homarus americanus* meet a larger one, the former will back down, but should two equal-sized beasts square up to each other they avoid a claw-ripping battle with some unusual behaviour – they urinate in each other's faces. The urine is stored in the bladder and can be released at will, and contained in it is chemical information about the status and health of the individual. Self-confident males tend to release urine quicker than less confident ones, in an attempt to encourage their opponents to retreat, but should the urine battle continue both squirt copious quantities of urine until one can be declared the winner.

It's thought that the loser will 'remember' the smell of the winner and avoid any confrontation in future.

Young horned ghost crabs *Ocypode ceratophthalma* change their appearance to remain camouflaged both day and night. They live in burrows on beaches in the Indo-Pacific region, from Japan to East Africa, but when they venture outside they become lighter by day to blend in with the sand and are darker at night. Their camouflage system is so well developed they can match the grains of sand on the beach on which they live and are less conspicuous to predators, such as seabirds and beach-combing monkeys. Fiddler crabs *Uca*, by contrast, tend to be darker by day and lighter in the night, an adaptation that either helps them maintain a comfortable body temperature or maybe acts as protection against harmful UV radiation.

Fiddler crabs 'talk' with gestures. They live on intertidal mudflats and among mangroves, the males instantly recognised by their one gigantic claw, which is waved about to threaten rivals and attract potential partners, but an overenthusiastic crab can get his message mixed up, attracting rivals while scaring away potential mates. The milky fiddler crab *Uca lactea* is subtler. He has a language of at least four signals: if a neighbour should emerge from his burrow and come too close, the resident will greet him with a series of short, sharp, side-to-side karate chops in order to resolve the potential border dispute. Distant passers-by are waved at with a large circular motion of the claw, thought to encourage other males to keep on walking but to attract passing females in a display that indicates the male's fitness. If a potential mate should approach more closely the male points at himself in a grand arching gesture, while bobbing up and down or running about. However, if a rival should approach, the resident squares up to him and waves his claw up and down right in front of his eyes. He's saying put up or shut up, and if he doesn't back down the two will come to blows.

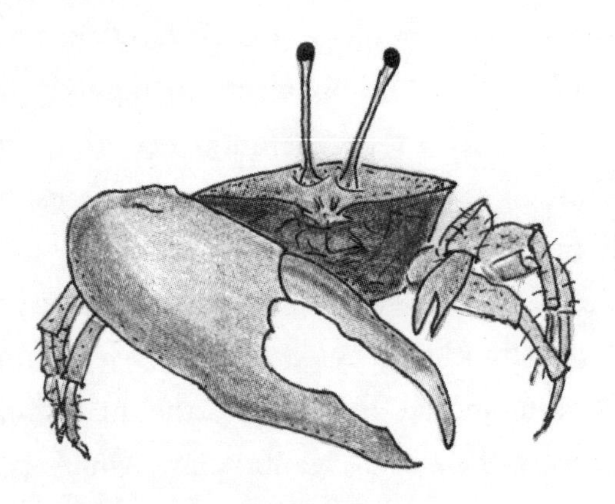

Fiddler crab

The male porcelain fiddler crab *U. annulipes* must be one of the most accomplished con artists on Earth. Like other fiddler crabs he waves his big claw about to threaten rivals and to attract potential partners; the bigger and more powerful the claw the better the male's ability to stand his ground or catch the attention of females. Sometimes, a male can lose his large claw, but it's only a temporary blip for he simply grows a new one. There is, however, a drawback. The new claw can be as big as the old one but it has less muscle, so if two crabs came to blows the male with the replacement claw would be at a disadvantage. Even so, this doesn't stop him waving it about. He pretends that everything is fine; in fact, the substitute claw is lighter than his original one and much easier to move, so he uses bluff to impress. Females seem unconcerned for they choose males with the longest claws, whether they are real or fake, and rivals are intimidated sufficiently to avoid a fight. His sham generally works.

Swimming crabs are smarter than your average crab. The East African swimming crab *Thalamita crenata* is active when the tide is going in or out, and hides in one of its many burrows at high and low tide. To navigate its way around its piece of beach and even negotiate short cuts, it recognises and remembers landmarks. Italian researchers know this because they changed around landmarks in order to fool the crab, but even though it was confused for a short time, it quickly learned the positions of the new landmarks and found its way home. This is something rare in invertebrates and previously unknown in crustaceans, and is comparable to route finding in honeybees.

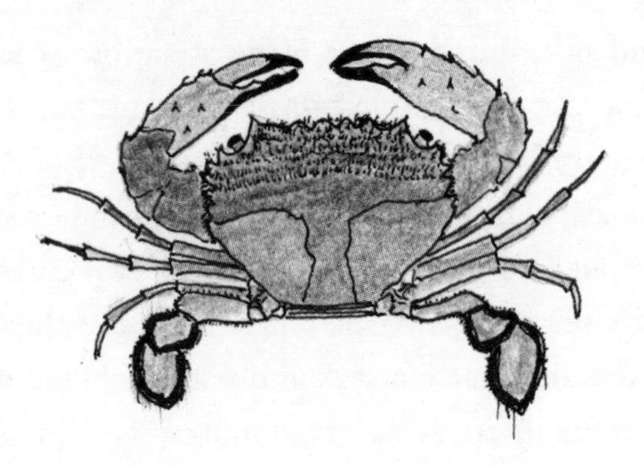

Swimming crab

Juvenile longnose spider crabs *Libinia dubia*, which live on the east coast of the USA, are so-called 'decorator crabs' and they do just what that suggests: they decorate their bodies with bits of seaweed to camouflage themselves from predators. But they don't pick up any old seaweed; they're very choosy. Even though other seaweeds are more common and easier to gather, they opt for a particular species of brown seaweed *Dictyota menstrualis*, which contains a bitter-tasting chemical – dictyol E – that is repulsive to fish. Juveniles continue to decorate themselves with the seaweed until they're fully grown and are too big for fish to tackle, and then they stop.

The big-clawed snapping shrimp *Alpheus heterochaelis* has one claw (which can be left or right) significantly bigger than the other. It lives in the tropical and subtropical western Atlantic and Gulf of Mexico where it and others like it create the most astonishing noise. It can snap its large claw to make a very loud bang. It was thought at one time that the sound was caused by the mechanical clacking together of the two moveable parts of the claw, but thanks to high-speed video Dutch and German researchers found that the sound is produced by an imploding cavitating bubble and spark in the high-speed water jet created as the claw snaps shut. The sound deters predators and is sufficiently powerful to zap prey, as well as being used as a display signal between shrimps.

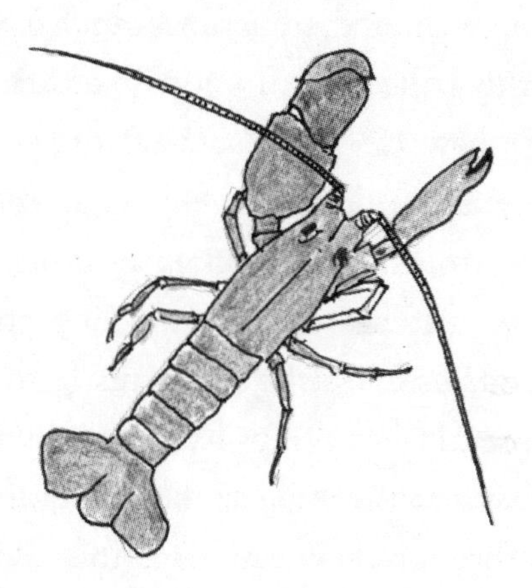

Big-clawed snapping shrimp

ON THE ROCKS

The bootlace worm *Lineus longissimus* is probably the longest animal in the world. It is claimed that a bootlace worm washed up on a beach near St Andrews, Scotland, in 1864 was 55m (180ft) long and only 10cm (3.9in.) wide. More usually specimens up to 9m (30ft) long can be encountered, sometimes tangled with seaweed or in an untidy collection of knots in a tidal pool, but if you find one, beware! This member of the ribbon worm phylum (Nemertea) oozes copious quantities of mucus laced with a strong neurotoxin as a defence. This voracious predator catches its own food by pushing out a proboscis and, using sticky filaments, effectively glues itself to its prey. It then wraps its slender body around to subdue it, like a constricting snake, and eventually sucks the food in through its mouth. It is found on European coasts in the north-east Atlantic.

Bootlace worm

On the California coast is the giant California sea cucumber *Parastichopus californicus*. It has a soft cylindrical, reddish-brown body about 50cm (20in.) long and 5cm (2in.) wide with the mouth at one end and the anus at the other. When threatened by a predator it squirts out its guts through its anus as a diversionary tactic, but how does it then feed and survive? Well, it feeds through its anus! Sea cucumbers have their respiratory system linked to the anus, where they suck in water to extract the dissolved oxygen. University of Washington researchers found that algae labelled with radioactive carbon ended up in the respiratory system. It is not clear how this food is digested and absorbed, but it is possible anal feeding might be a way of taking in extra food, which can also be utilised while the sea cucumber's gut is growing back.

Starfish eat mussels but mussels are sometimes on parts of a rocky shore exposed at low tide – so how can a hungry starfish keep on top of its prey and not overheat or dry out? The purple or ochre sea star *Pisaster ochraceus*, which lives on Pacific coasts, is one species with a solution. If it finds itself marooned, it sits out the discomfort and waits for high tide. It then absorbs water so that at the next low tide it can remain cool and not overheat, even though it might be exposed to the sun. It will continue its water trick until it moves off into deeper water.

The microscopic, free-floating larva of a sea urchin is not an especially bright creature, but it has to make one major decision in its early life: it has to settle down on a rocky shore and develop into the adult form. Its parents have cast it out there in the open sea, at the mercy of the ocean currents, but how does it know if it has reached the right neighbourhood? Researchers at the Bodega Marine Laboratory have worked out how it knows that the time is right. Their study animal is the purple sea urchin *Strongylocentrotus purpuratus*, a common species living on the Pacific coast of North America. It is one of the giant kelp forest community, along with abalone and sea otters. Researchers discovered that the larvae have a two-step process to settling. Firstly, they respond to the shearing forces when tumbled by waves close to a rocky shore. This alerts them that they are in roughly the right place. Secondly, they are sensitive to chemical traces in the water produced by the thick carpets of algae on which the adults feed. At this point they drop to the seabed and develop into spiky juvenile sea urchins – job done!

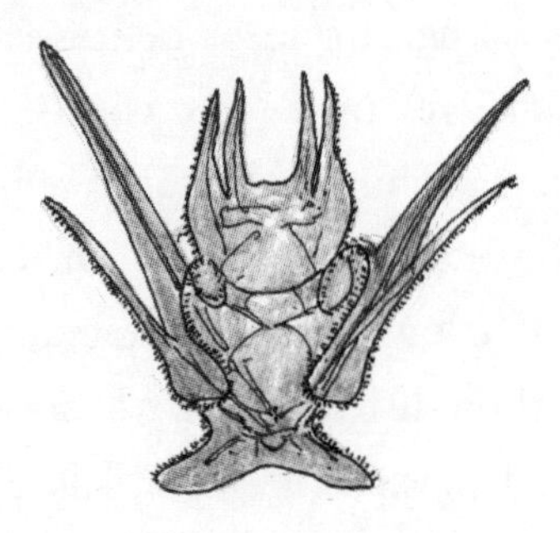

Sea urchin larva

When you look at a sea urchin, 'vision' is not the first sense that comes to mind, after all it doesn't seem to have any eyes, yet the purple sea urchin can see a large object, appreciate its distance away and move away or towards it. It is thought that it can 'see' things with photoreceptors in its feet and by channelling light down its spines to light-sensitive skin cells that form a kind of 'retina' below the test (shell); and purple sea urchins have the most spines of all known species of sea urchins. It means that the sea urchin's body is like a single enormous compound eye, and it is thought that vision in this species is as good as that of a horseshoe crab, which does have eyes.

Sea hares *Aplysia* are molluscs. They look like wavy slugs with bunny ears, which are actually 'rhinophores' or smell organs, and as slow-moving, soft-bodied creatures they look extremely vulnerable, but there you would be wrong. They tend to blend in with their background, taking on the colour of the seaweed on which they browse, and their skin is laced with noxious chemicals sequestered from the algae. But when push comes to shove and a predator, such as a lobster, comes a little too close the sea hare releases a cloud of ink, its colour – white, purple or red – dependent on the seaweed it has been eating. It is not just a smoke screen, though. The milky cocktail is the product of two glands – the ink gland and opaline gland – and it contains a chemical that gums up the chemical sensory system of its aggressor.

It effectively 'blinds' the predator, literally knocking it chemically senseless.

A sea hare and lobster

The parasitic fluke *Microphallus piriformes* infests two hosts – firstly the rough periwinkle *Littorina saxatilis* that lives among the seaweeds on rocky shores, and secondly the herring gull *Larus argentatus* – and the fluke has a clever way to get from one host to the other. When mature, it changes the behaviour of the periwinkle so that instead of hiding, the mollusc follows the rising tide upwards so it is exposed on the tops of rocks – just the place where herring gulls can spot it. As gulls are the ultimate opportunists, they immediately eat it and the parasite is passed rapidly from one host to the next.

The predatory dog whelk *Nucella lapillus* often targets the common mussel *Mytilus edulis*. It uses its tooth-like radula to drill into the mussel's shell in order to get at the soft parts inside. The mussel, however, has a way to fight back. It usually anchors its shell to the rocks by secreting tough anchor lines, known as byssus threads, but when a dog whelk attacks it can throw out more of these threads and entangle the attacker, stopping it in its tracks. University of Dublin researchers watched mussels doing just this. They witnessed a group of five mussels immobilising a dog whelk with forty-two byssus threads.

Living on a rocky shore can be turbulent for marine creatures and one species, the northern clingfish *Gobiesox maeandricus*, has found a way to hunker down and not budge. It has an adhesive pad on its underside that is edged with tiny hairs (microvilli), similar to the spatula-shaped bristles (setae) on the feet of geckoes. These help it stick to the slippery and irregular surfaces of rocks with a suction that out-performs man-made suction cups, which can only adhere to smooth surfaces.

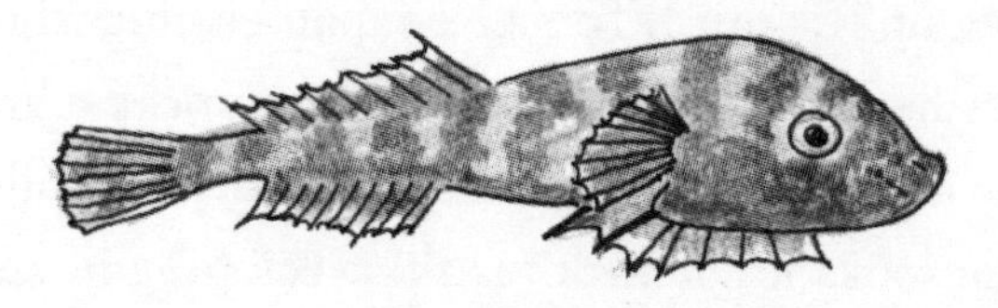

Northern clingfish

MUD, MUD, GLORIOUS MUD

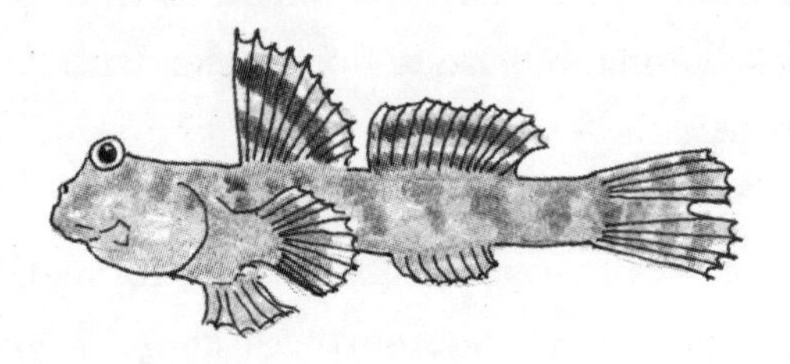

Mudskipper

The mudskipper is a curious fish. It lives in mangrove swamps and on mudflats at the edge of tropical and subtropical seas, and it is just as active out of water as in. As long as it keeps moist it can breathe through its skin, much as amphibians do. It also has an intriguing way of bringing up its young. The male Japanese mudskipper or Shuttles hoppfish *Periophthalmus modestus* excavates a burrow in the mud in which the female deposits her eggs. He then looks after them for the week-long incubation period but in an extraordinary way. The eggs develop in air, where the temperature is higher, there is more readily available oxygen and they are safer from marine predators, but they hatch in water. The nest burrow, however, is permanently underwater so how can the eggs develop in air? The answer is surprisingly simple. The male mudskipper emerges at low tide, swims to the surface and takes a mouthful of air, releasing it when he returns to the nest chamber. It does this several times until there is sufficient air in the chamber to last until the tide goes out again, when he can repeat the exercise. At the end of the week, he makes sure his youngsters hatch in water by reversing the process. He gulps air from inside and releases it outside, so water flows in.

Certain types of clams 'mine' the mud with their feet and in doing so stretch their bodies more than any other creature on earth. They are in the family Thyasiridae, and they harbour colonies of bacteria in their tissues that require certain chemicals – sulphides – to manufacture food. It's up to the clam to find them and the bacteria to turn them into sugars with which the clam supplements its normal diet of filtered food particles. The sediments are without oxygen and sometimes deep beneath the muddy seabed, but the clam has a way to reach them. It pumps body fluids into its single foot and prevents them from returning to the body so the foot slowly elongates, a bit like an earthworm burrowing in the soil. It can stretch its foot up to thirty times its relaxed length.

On the tidal mudflats of Baja California, the male whelk *Solenosteira macrospira* looks after its annual brood, rather than the female, but not all the eggs are his own. Rampant promiscuity on the part of female whelks means that he is well and truly cuckolded and can be caring for the eggs of more than a dozen other males! But that is not all: the entire reproductive process is unusual. After the snails mate, the female glues tiny capsules, each containing hundreds of eggs, to the male's shell. He becomes a substitute rock since the mudflats offer few places to attach eggs, and moving with the tide on father's or stepfather's back ensures they are not exposed to drying out and the extremes of temperature that they would have experienced on a stationary rock. Even so, only one in four eggs is his. After about a month the youngsters are ready to hatch out, but the intrigue does not end there: a few dominant babies devour the rest of their littermates, so only about half a dozen survivors emerge from the capsules and crawl away.

The spoon worm *Maxmuelleria lankesteri* is an inconspicuous green, sausage-shaped marine worm with a tube-like proboscis with which it grazes the sea floor. It can be found off the west coast of Scotland, where it lives in a deep burrow in gloopy mud ... at least females do, for these worms show extreme size difference between the sexes. The female may be up to 30cm (12in.) long, but the male is no bigger than a few millimetres and lives permanently as a parasite inside the female's body.

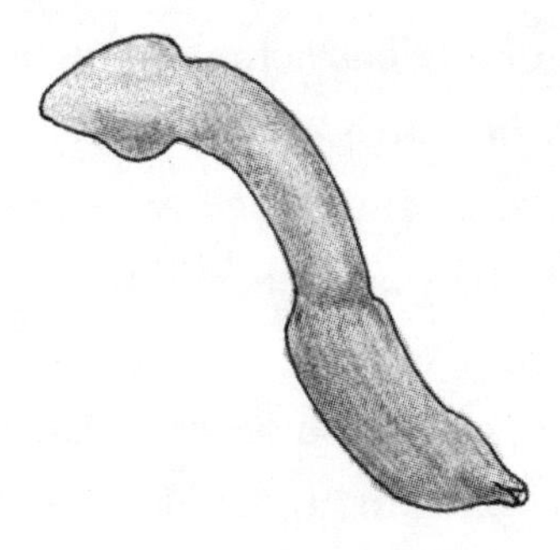

Spoon worm

Another species, the fat innkeeper worm *Urechis caupo* has a unique way of gathering food and keeping its house tidy. It traps tiny food particles with a conical, mucous net set in a large U-shaped burrow through which the female worm maintains a steady flow of water. By pulsating her fat pink body she moves about 18 litres (32 pints) per hour. And when it comes to getting rid of waste she has a simple solution. She accumulates faecal pellets and burrow material at the end of her tunnel and periodically she squeezes her body violently, which sends a jet of water through her anus to blast out the accumulated debris. She can be up to 50cm (20in.) long and lives on the west coast of North America, and she might not be alone. She can share her burrow with pea crabs *Pinnixa* and *Scleroplax*, scaleworms *Hesperonoe*, clams *Cryptomya* and gobies *Clevelandia*.

Deep-sea relatives of spoon worms can be up to 8m (26ft) long and although they have been seen from submersibles, nobody has actually caught one. In Korea spoon worms, which are known locally as 'penis fish', are considered a delicacy and eaten raw with salt and sesame oil, and in China they are stir-fried with vegetables.

When it comes to gruesomeness Nature wins hands down. Take the parasitic barnacle *Loxothylacus panopaei*, for example, which lives as a featureless carbuncle on the estuarine mud crab *Rhithropanopeus harrisii*. When conditions are right – a sea temperature of 25°C (77°F) and salinity of 20 per cent – it releases broods of larvae every five or six days that float around in the sea. Each successful larva eventually grabs on to an adult or juvenile crab just after it has moulted and injects a wriggling, worm-shaped bag of cells into the body of its victim. The bag bursts and the cells disperse via the crab's blood system by amoeba-like movements. Any one of the cells (and sometimes more than one) can develop into the adult parasite – a real-life *Alien*.

CORAL REEF CHARACTERS

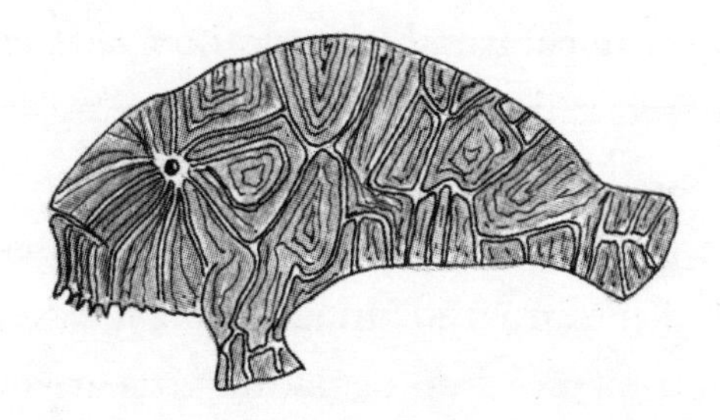

Psychedelic frogfish

Is it a bird? Is it a plane? No … it's a psychedelic frogfish *Histiophryne psychedelica*. This striking, if somewhat bizarre, coral reef fish was discovered on reefs around the Indonesian islands of Bali and Ambon by University of Washington biologists. Its most striking feature is its colour – a swirling pattern of orange, yellow and white stripes radiating from the position of each turquoise-ringed eye and continuing over the rest of the body. It closely resembles the corals among which it hides and is therefore almost invisible. Its fins resemble feet, with which it clings to the reef or crawls into crevices, where it can plug the entrance with its flabby-skinned body and trap prey inside. If a target is located it pushes out its jaws, expands its mouth up to twelve times its resting size, and grabs its victim ... and all in just six milliseconds. The frogfish can swim, but not like other fish. It uses jet propulsion, like a squid. It takes water into its mouth and forces it out through its gills in short bursts that see it leapfrogging along the seabed.

Clownfish or anemonefish *Amphiprion* hide from their predators among the stinging tentacles of sea anemones, but what does the anemone receive in return? Scientists at Auburn University, Alabama, USA, have discovered at least part of the answer. During the night, when oxygen levels on a coral reef drop significantly, the movements of the clownfish among the anemone's tentacles boost oxygen flow. The little fish have three behaviours – fanning, wedging and switching, the equivalent of tossing and turning in bed. When fanning, the fish remain motionless and move their fins rhythmically. While wedging, they wriggle deeper into the bed of tentacles, and during switching they change their orientation; in fact, they're active for 80 per cent of the night. The health of their hosts is far more important than drawing unwelcome attention to themselves. Nemo, it seems, is an insomniac!

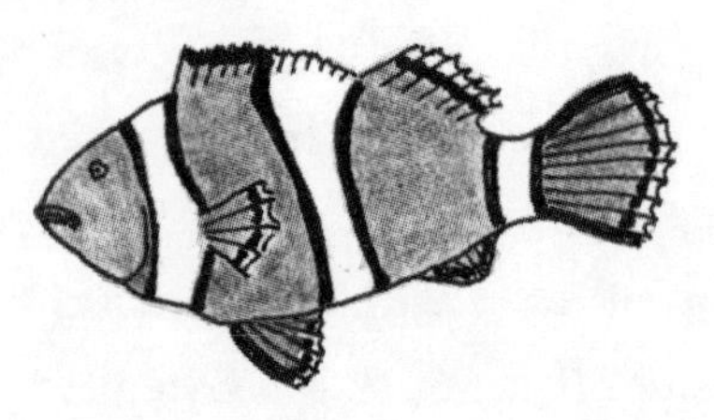

Clownfish

Nemo is also into gender reassignment. When several clownfish share an anemone there is a hierarchy with a single aggressive female the dominant fish. The rest are males, and the female will breed only with the most dominant male. If she should die, the dominant male changes sex to become the new female, and all the other males move up a step in the pecking order.

Male and female sea ponies *Hippocampus fuscus*, a species of seahorse, dance at dawn. They live in sea grass beds *Zostera* in Sri Lanka and greet each other every morning by changing colour and performing a little dance lasting four to five minutes. Then they go their separate ways for the rest of the day. The male, rather than the female, carries their fertilised eggs in a special pouch and he gives birth eventually to tiny versions of the adult seahorse. At this point, the morning dance is prolonged, lasting up to nine hours, after which they mate again, for seahorses are partners for life.

Sea ponies

Roving coralgroupers *Plectropomus pessuliferus* in the Red Sea team up with giant moray eels *Gymnothorax javanicus* to better their chances of catching a meal. The groupers, which are diurnal, encourage the normally nocturnal eels to join the hunt by shaking their heads. The grouper chases the fish into the coral and the moray eel flushes them out, straight into the jaws of one of the partners. Whoever is first swallows the fish whole so there is no fight; in fact, over the course of a fishing expedition both the grouper and the eel catch fish making the cooperation worthwhile. It even transpires that the grouper appears to point deliberately with its head towards the target. It will hover over the place where the prey is hiding and make a 'headstand signal', tilting its body so its head is pointing down and then shaking it in the direction of the potential food. Coral trout *Plectropomus leopardus* on the Great Barrier Reef perform a similar signal to big blue octopuses *Octopus cyanea* in order to maximise their chances of catching food. Such cooperation between species is extremely rare and usually confined to animals with higher brain processing power than that of a fish, and until now pointing gestures were the preserve of primates and one other animal – the raven.

The grey reef shark *Carcharhinus amblyrhynchos* is one of the more common species of shark to be encountered on the outer edge of coral reefs and atolls. It usually patrols the drop-off into deep water, where the currents are strong and food is plentiful, and it can be aggressive if approached too closely. It has gained a respectful notoriety because of its characteristic behaviour towards divers. If approached too closely it gives a warning: it drops its pectoral fins, arches its back and swims stiffly, and if the threat doesn't back down it attacks with a slashing bite.

Grey reef shark

The sharks are found on reefs in the Indo-Pacific region, but surprisingly little has been known about them; that is, until new research on mostly female sharks by a consortium of Australian universities has revealed that the moon, the sun and the seasons govern their daily patterns of behaviour.

The sharks were observed on reefs at Palau in Micronesia, where many females spend much of their lives swimming around the same general sites, but change their position in the water column both horizontally and vertically on a daily and seasonal basis. In winter they remain close to the surface at a depth of about 35m (115ft), where the water is warmer. In spring, when the sea starts to hot up, they dive to cooler waters at 60m (197ft) deep.

The sun also determines a daily vertical migration. They move into shallower water, at an average depth of 30m (98ft) at dawn, hitting a depth of about 45m (148ft) by midday, before drifting back towards the surface in the afternoon and into shallower water by dusk.

At night they also change their behaviour in time with the lunar cycle, diving much deeper during the full moon, when the moon is at its brightest, and shallower with the new moon. These behaviour patterns are thought to help the sharks conserve energy and find food, their daily vertical migration following that of their prey.

Whitetip reef shark

Many smaller coral reef fish hide at night for fear of being caught by marauding gangs of whitetip reef sharks *Triaenodon obesus*. They have wedge-shaped heads which they push into cracks and crevices so fish sleeping there must make themselves as inconspicuous as possible. Several species of parrotfish, including the queen parrotfish *Scarus vetula*, take concealment to a new level. When darkness envelops the reef the parrotfish starts to make a sleeping bag. Long gossamer threads of mucus exude from its mouth and by moving its body it weaves a mucous envelope that covers its entire body, save for a flap with a small hole for its mouth and another behind its tail that enable the flow of water to continue over its gills. It is thought the 'sleeping bag' prevents most of the fish's scent from escaping and giving away its location. It might also offer an early warning if a shark or moray eel should come calling and touch the bag.

Parrotfish

At night, some species of pearlfish, e.g. the star pearlfish *Carapus mourlani*, hide inside the bodies of sea cucumbers, and they lodge in the most unlikely place – the sea cucumber's anus. For most of the day they forage in the open, but come evening they search for somewhere to spend the night. They first inspect a prospective host, avoiding the head end, and, guided by smell, make for the other end. They gain entry through the anal pore by wriggling through either head first or more frequently tail first. There can often be two fish inside each sea cucumber – a male and female – but there can be more. One large leopard sea cucumber *Bohadschia argus*, which was 40cm (16in.) long, was found to have fifteen pearlfish inside, the highest number ever reported from inside a sea cucumber. The pearlfish also lodges inside large starfish, especially cushion stars, such as *Culcita novaeguineae*, which resembles an inflated pentagonal pin-cushion about 30cm (12in.) across.

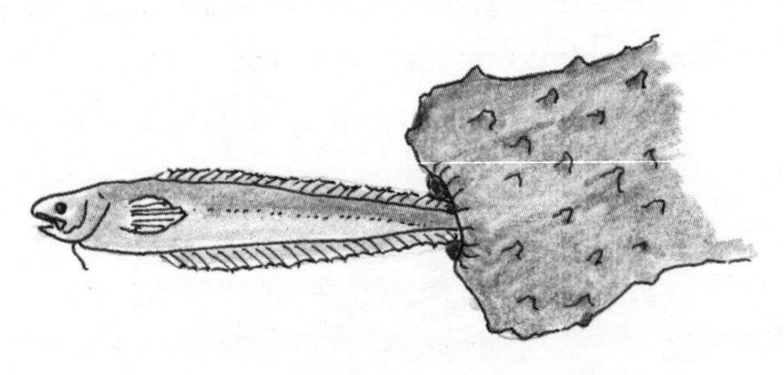

Pearlfish hiding inside a sea cucumber's anus

Living coral is always at war with algae, overgrowth being a major threat, but corals have novel ways to keep it at bay. In Fiji, for example, emerald-green mats of turtle weed *Chlorodesmis fastigiata* threaten to smother a small species of staghorn coral *Acropora nasuta*, one of the main reef-building corals. In response, the coral sends out a chemical distress signal that is picked up by broad-barred gobies *Gobiodon histrio* and redhead gobies *Paragobiodon echinocephalus*, small fish that trim the turtle weed. The gobies, however, will only eat the weed if the coral summons them, even though they actually gain protection themselves. When eating the weed, the gobies' skin secretions are more toxic and therefore more likely to repel predators. And there is one more advantage: in return for keeping the weed away, the coral provides the fish with a safe shelter.

The tiny sea-anemone-like polyps of some types of soft corals, e.g. *Heteroxenia fuscescens*, pulsate. Their feathery tentacles are constantly in motion, behaviour that was first recognised by Jean-Baptiste Lamarck over 200 years ago, and something that only their free-living jellyfish relatives do. At first, it seemed that the movements were for feeding, but several studies have shown that these corals do not feed by predation. Even so, the movements are costly in energy terms, so there must be a good reason to do it. Scientists at the Hebrew University of Jerusalem think they know why.

The first discovery was that the water around the coral is swept up and away, meaning that water is not stagnating and re-filtered time and again. Fresh, incoming water, rich in essential gases and nutrients, constantly bathes the polyps. These corals rely on algae living in their tissues for their food, and the researchers found that the pulsating tentacles act like a fan, ensuring that waste oxygen from photosynthesis is wafted away so the algae are afforded the best conditions for optimum food production. In fact, photosynthesis is ten times higher when the polyps pulsate than when at rest. The second surprise was that the polyps take a siesta. They stop pulsating in the afternoon.

HEADS OR TAILS?

Sea snakes have heads on their tails and tails that 'see'. The yellow-lipped sea krait *Laticauda colubrina* appears to have two heads. When foraging around coral reefs with its head in a crevice, the tail is twisted so that it looks like the top of the head. This false head ensures predators, such as sharks, are deterred from attacking for they would not welcome a bite from an animal that has one of the most potent venoms on Earth.

Yellow-lipped sea krait

The reef-dwelling olive sea snake *Aipysurus laevis* has photoreceptors in its tail. It hides among coral clumps for much of the day, but a telltale tail sticking out of the corals would alert a predator, such as an osprey flying overhead, to the snake's presence. It is thought the light-sensitive tail tells the snake if it is fully concealed inside its daytime coral cave home.

The marine betta or comet *Calloplesiops altivelis* is a dark, spotty fish, about 20cm (8in.) long, which lives on tropical coral reefs. Being nocturnal, it spends the day hidden in crevices on the reef, and it has a clever trick to ensure it's not disturbed. At the base of its dorsal fin is a large eyespot. The comet puts its head end into the crevice but leaves its back end sticking out of the hole. It also expands its caudal, anal and dorsal fins, exposing its eyespot fully, and the gap between its dorsal and anal fins resembles an open mouth. With this it looks remarkably like an aggressive white-spotted turkey moray *Gymnothorax meleagris* poking its head *out* of the crevice, so potential predators leave it well alone. It can also use its eyespot when it goes hunting. It moves sideways towards its target fish, and waits for it to try to escape. There is a 50–50 chance that the fish will dart towards what it perceives as the tail end, which is actually the comet's mouth.

Marine betta and white-spotted moray

CORAL REEF VALET SERVICE

Cleaner wrasse and grouper

Fish say 'sorry'. The cleaner wrasse *Labroides dimidiatus* feeds on external parasites and dead or infected skin and scales, providing a cleaning service for larger fish that stand in line to be spruced up. The wrasse calms the potential predator by stroking the client's body with its fins, and will use fin stroking to extend a cleaning session even if the client is attempting to move away. Predatory clients receive more fin-stroking than non-predatory ones in an attempt to quell their killer instinct, and should a cleaner accidentally hurt a client it will follow up with plenty of touchy-feely behaviour to appease the larger fish. It is the first time 'apologising' has been observed in a non-mammalian liaison, which might challenge the notion that larger brains have developed to deal with complex social arrangements. Cleaner wrasses maintain customer-client relationships with perhaps 100 different individuals from fifty or more species. It begs the question, why do primates need big brains to handle their social lives?

You would think cleaner gobies *Elacatinus* would rush to offer their services to the non-threatening algae-eating fish, which are unlikely to harm the cleaners, rather than predatory fish, which could eat them, but you would be wrong. Like the cleaner wrasse, the gobies react more quickly and provide a higher level of service to predators. It is thought the gobies choose the risky customer over the harmless ones to identify themselves immediately as service providers rather than the predator's dinner. Unfortunately, predators such as groupers scare away the rest of the cleaners' clientele, so they might well be speeding up their service just to get rid of the dangerous fish as quickly as they can so that things can get back to normal.

Similarly, cleaner shrimps *Periclimenes longicarpus* clap their claws together to attract clients to their wash-and-brush-up service. These delicate, partly transparent but colourful crustaceans hide among the tentacles of sea anemones but emerge to divest reef fish of their parasites and dead skin. It is at this time, away from their protective host, that they are vulnerable to being gobbled up by passing predators. They try to discourage an attack by clacking their pincers even more vigorously, an attempt it is thought to remind those that wish to do them harm that they have a valuable service on offer. The frenetic clacking is a signal that might save their lives – pre-conflict management.

FISH TALES

Early summer in the North Atlantic can be extremely noisy. Male Atlantic cod *Gadus morhua* grunt on their breeding grounds, and many more 'white fish' squeak, burp and groan to add to the chorus. The cod congregate in traditional spawning areas, always returning to the same site, where they form vertical stacks, known as 'haystacks', while waiting for the cod of their dreams. Low-frequency grunting mainly occurs during the day from late May to early June, the sounds thought to be attractive to females but also to warn off other males.

Most fish (and most vertebrates) have red blood (when infused with oxygen), while crustaceans and molluscs have blue blood, but crocodile icefish (Family: Channichthyidae) have blood as clear as a glass of gin. Most icefish live in relatively shallow waters, at depths less than 300m (984ft), but one species – the 50cm (20in.) long ocellated icefish *Chionodraco rastrospinosus* – lives in the deep, up to a kilometre (3,280ft) below the surface of the Southern Ocean. How it and its semi-translucent relatives survive without red blood corpuscles is puzzling. In normal fish, about 90 per cent of oxygen is carried around the body by the red blood cells, with the remaining 10 per cent dissolved in the blood plasma. Icefish must adopt another system. They have no scales, so a small amount of oxygen could be absorbed through the skin; after all, colder waters contain a far greater concentration of oxygen than warmer seas. The fish have a larger than normal heart (three times bigger than the hearts of other fish of the same size) and wider blood vessels so the blood is pumped around the body in greater volume (they have four times more blood than other fish) and more rapidly, with blood plasma carrying most of the oxygen. They are also rather sluggish, resting on the seabed for much of the time, supported by their pectoral fins, so their need for oxygen is less than in more active fishes. Another odd thing is that the icefish have natural antifreeze in their blood, an adaptation that prevents their body fluids from freezing at temperatures down to –2°C (28°F), enabling them to survive in Antarctica's ice-cold waters.

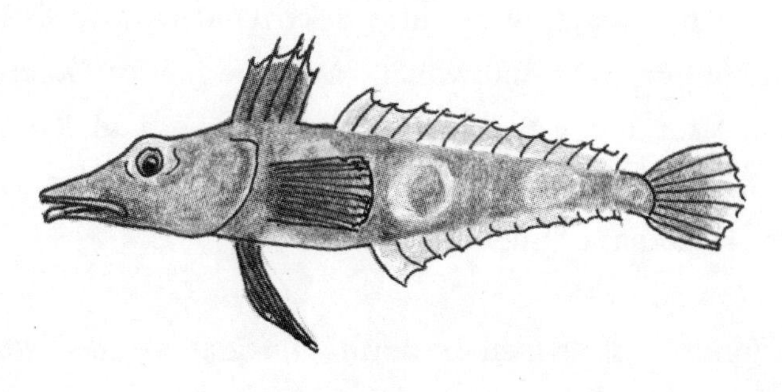

Icefish

During the breeding season in the western Pacific, the male long-finned or sleeper striped goby *Valenciennea longipinnis* occupies a burrow in which the female has deposited her eggs. He remains with the eggs, guarding them against nest predators and fanning them with water to ensure they develop under optimum conditions with high levels of oxygen. The female, meanwhile, builds a mound of stones and coral debris outside the burrow entrance, and she maintains its height despite the action of waves and currents. The mound, it turns out, is vital for a successful brood, for it causes a pressure gradient between the burrow openings which helps water flow inside the burrow. If the mound should deteriorate, the male must fan faster, and if the sad state of repair continues he will eventually abandon the nest.

After spending the summer away, the island or comb grouper *Mycteroperca fusca* is probably glad to be back home. In a study of these groupers on the coast of Madeira in the eastern Atlantic, which lasted for twenty-five years, researchers discovered that an easily recognised yellow individual leaves the Bay of Garajau on the south coast of the island in spring, and returns to the exact same spot in autumn, an annual migration similar to that of migratory birds.

A handful of striped beakfish *Oplegnathus fasciatus*, which normally live on the reefs around Japan and

Hawaii, made an unusual journey. They travelled across the Pacific Ocean in the submerged stern compartment of a 5.5m (18ft) fibreglass skiff, a victim of the Japanese tsunami of March 2011. The fish, along with crabs, scallops, blue mussels, algae, marine worms, sea anemones and a sea cucumber – all told up to fifty species of sea creatures – found refuge in the cave-like lidless compartment and were eventually driven ashore at Long Beach in Washington state on 22 March 2013. It alerted the US authorities to a new way in which invasive species could enter North American waters.

BIRDS AT SEA

Swallow-tailed gulls *Creagrus furcatus* living on the Galapagos Islands do not hunt or scavenge during the day, but fly at night. They're the world's only species of nocturnal gull, and their pattern of hunting at sea is linked to the phases of the moon. On nights with a new moon, during which the moon is practically invisible, their prey – zooplankton, small fish and squid – rises to the surface under cover of darkness to feed, returning to the depths to hide during the day. This is when swallow-tailed gulls are on the wing. Their eyes are large and well adapted to the dark, with a shiny tapetum that reflects light back through the retina, enabling them to see at very low light levels. They can spot fish just below the surface without any light from the moon. However, on bright moonlit nights, such as at full moon, the gulls don't bother to go hunting as their prey will fail to show.

The king penguin *Aptenodytes patagonicus* dives down to a depth of 300m (984ft) in less than seventy seconds, during which time its eyes must adapt quickly from bright sunlight to twilight in order that it can see its prey, mainly lanternfish less than a centimetre long; yet it can catch

2,000 fish in twenty-four hours in dives that last four to five minutes. It achieves this with some remarkable eyes. It is the only bird that can shrink its pupils down to tiny square-shaped pinholes, minimising the amount of light that enters while the bird is on land, while on a dive it can expand them to wide, circular apertures about 13mm (0.5in.) across, allowing a 300-fold increase in light to enter the eyes. In the deep it can pick up the faint light from the light organs of the lanternfish (Family: Myctophidae), exploiting a source of food denied to other seabirds because they simply can't see them – but the king penguin can.

Life must be tough for the little guys, and life couldn't be tougher for the smallest of the auks – the little auk or dovekie *Alle alle*. It spends much of its life in the icy cold Arctic, but in the winter it heads south to warmer temperate waters, such as those off the New England coast where it falls prey to an unexpected predator – the monkfish, known locally as the goosefish *Lophius americanus*. The monkfish, however, normally lurks on the seabed where it entices prey close to its mouth care of a dorsal fin spine with a lure that is waved about like a worm. So how does a bottom-dwelling predator gobble up a small seabird, which cannot possibly reach the sea floor in this deep part of the ocean? Monkfish, it seems, leave the confines of the seabed and swim up towards the surface when it is time to spawn or during offshore-onshore migrations, when they rise to within 6–9m (20–30ft) of the surface and take advantage of favourable ocean currents. When the monkfish are on their way up, the little auks are on their way down, so what is a self-respecting monkfish going to do? It grabs an auk, of course, in its enormous mouth. And monkfish are not the only predators – spiny dogfish, Atlantic cod, red hake and red-spot flounder will take the birds too.

Petrels and related seabirds (Order: Procellariiformes) have enormous nostrils and it was long thought that they used the sense of smell to find their food, though there was no proof. It now turns out that this is probably correct, due to some field experiments by researchers from the University of Washington. They put slicks of vegetable oil laced with dimethyl sulphide (DMS) into water and watched what happened. Sure enough, several species of petrels and prions found the slicks even in the dark. The significance of DMS is that this is the chemical given off by phytoplankton when attacked by zooplankton, such as krill, usually at night, and it's the krill that the petrels want to catch. Using their sense of smell they locate the besieged phytoplankton and thus the swarms of krill and are rewarded with a midnight feast. But the story doesn't end there. A later experiment using extract of the krill itself mixed with the vegetable oil slick also attracted petrels, but failed to impress other seabirds. It means that the olfactory world of these birds is more complex than simply seeking out a single scent. It also means that these birds could be using olfactory cues, such as areas of high productivity with phytoplankton blooms and swarms of krill, in navigating their way around the otherwise featureless ocean.

A small colony of Laysan albatross *Phoebastria immutabilis* on the Hawaiian island of Oahu has a chronic shortage of males, so some of the females there have opted for an unexpected solution. Out of the forty nesting pairs, fifteen comprise pairs of females, one of which lays two eggs. The absent fathers are already paired with their own females – a partnership for life – but they're not averse to a little canoodling with the unpaired females. The 'gay' couple, however, bring up a single chick (for the other egg doesn't hatch) as normal, each sharing the incubation and food-fetching duties, and the relationship can be enduring: one same-sex pair has been together for nineteen years. Same-sex relationships are actually common in nature with more than 450 species known to exhibit male–male or female–female relationships, although long-term relationships are rare. It's thought that in the albatross's case it's preferable to not breeding at all.

Animals in polar regions tend to have thick feathers or fur with an underlying layer of fat or blubber, but the European great cormorant *Phalacrocorax carbo* has neither of these. It is poorly insulated, yet it can survive the rigours of life in places like Greenland, where the Arctic winter is unforgiving. Instead, the cormorant has a behavioural trick up its sleeve. Research at the UK's Centre of Ecology and Hydrology compared cormorants at Disko Island in west Greenland, where the winter water temperature is about 5°C (41°F), with birds in Normandy where the sea temperature is a balmy 12°C (54°F). Surprisingly birds at both locations had the same body weight and ate the same amount of food, so how did the northern birds cope with their icy home? The answer is that the Greenland birds catch their fish quicker. They spend just forty minutes each day in the water, whereas their southern cousins were at sea for two and a half hours. A third group of birds at a fjord near Niaqornaasuk are even more efficient. They spend most of their winter sleeping on an icy sea cliff and enter the water, with an average winter temperature of –1°C (30°F), for just nine minutes each day. They catch fish at a rate thirty times higher than that recorded previously for any seabird.

PERFORMING SEALS

Ronan is undoubtedly cool. She's not a 'cool-cat', but a cool California sea lion *Zalophus californianus* who bobs her head in time with music. She was trained by researchers at the University of California, Santa Cruz, to move her head to a section of 'Down on the Corner' by Creedence Clearwater Revival, and when she was presented with 'Everybody' by the Backstreet Boys and 'Boogie Wonderland' by Earth, Wind and Fire she had no problem in keeping the beat even though she had not heard the songs before. Until this experiment, it was thought that only animals capable of vocal mimicry, such as humans, parrots and budgerigars, could 'keep time', but Ronan brings that assumption into question. It may also mean that there are many more music aficionados out there in the wild.

Harbour or common seals *Phoca vitulina* sport a splendid set of whiskers and they are turning out to be remarkable sense organs. Researchers at Germany's University of Rostock have discovered that they can detect the presence of a fish even though it had swum past thirty seconds previously. The whiskers detect the vortices (spiralling currents) left behind in the fish's wake. With this sense the seal can detect a meal from some distance away – up to 40m (131ft) – and, because different species of fish have different wake signatures, this wily predator can probably tell what is on the menu.

People remember voices, but can other animals? Well, the answer appears to be yes. A researcher from the Smithsonian Institution in Washington DC played archive recordings of mother and baby northern fur seals *Callorhinus ursinus* to a colony on St Paul in the Pribilof Islands and found that both mother and offspring could remember and respond to each other's calls four years after they were last together. This was the first time in a non-human mammal that long-term recognition had been demonstrated.

Size for size, mammalian carnivores with flippers, such as seals and sea lions, have significantly larger brains than carnivores with feet. It's thought that the demands of chasing prey which can move in three dimensions requires more brain power than hunting, say, on the open plains.

Grim scenes are being played out on Dundas Island, one of the subantarctic Auckland Islands, 320km (200mi) south of New Zealand. Male New Zealand or Hooker's sea lions *Phocarctos hookeri* abduct pups of their own species and carry them out to sea, where they shake them violently from side to side to kill them and then bite off the limbs and chunks of flesh and swallow the pieces. These scenes of cannibalism are not especially common, with just one in fifty pups killed this way. Why the males do it, however, has to be pure speculation, but it's thought the pups are easy targets and packed with energy-rich food, a single pup providing the daily needs of an adult sea lion.

WELL-TRAVELLED SEALS

The southern elephant seal *Mirounga leonina* is the world's largest living species of seal. Males can be up to 4.9m (16ft) long and weigh over 2,500kg (5,512lb), almost the weight of a large luxury car. They are also the deepest-diving seals, sometimes going down to depths of 2,000m (6,562ft) or more to hunt, resting and even sleeping on the way down and on the way up. They can hold their breath for up to 100 minutes, the longest of any non-cetacean sea mammal, but more usually they go down to depths of about 500m (1,640ft), staying underwater for about twenty minutes for females and an hour for males. But, once down in the eerie darkness of the deep sea, how do they find their prey? They do not have echolocation like the toothed whales and dolphins, but they do have large eyes equipped with plenty of light-sensitive rods, and research at the Chizé Centre for Biological Studies in France has uncovered how they use them. They hunt in parts of the deep sea that are illuminated by the faint blue glow of bioluminescent sea creatures, especially their main prey – the conspicuously bioluminescent lanternfish, the most numerous of all the deep-sea fish. Living light is their secret for a successful meal.

Elephant seals are known to roam great distances across the ocean between traditional breeding sites and feeding sites. Male northern elephant seals, for example, travel to Alaska from California to feed, and females may reach as far west as Hawaii. A southern elephant seal, nicknamed 'Jackson', tagged on a beach at Tierra del Fuego, was found to have travelled 28,970km (18,000mi) during the course of a year, but the all-time long-distance champion must be a male southern elephant seal that was born on subantarctic Macquarie Island in the south-west corner of the Pacific Ocean. He was found to be breeding successfully with his harem of females on a beach in the Falkland Islands in the South Atlantic on the opposite side of Antarctica.

TURTLE TRAVELS

Some green sea turtles *Chelonia mydas* that feed on the Brazilian coast do not breed there; instead, they head out into the open ocean and find tiny Ascension Island just 20km (12mi) across, which sits on the mid-Atlantic Ridge in the middle of the equatorial Atlantic. Why they travel the 2,300km (1,429mi) and not breed on perfectly adequate beaches in Brazil is a mystery and how they find such a small speck in a vast ocean is proving to be equally puzzling. It is speculated that the Earth's magnetic field plays a role, as does a sun compass during the day and the familiar smell of the place on the wind. Recent research indicates that non-magnetic cues are important for most of the journey with magnetic cues filling in the gaps, especially the last 50km (30mi), but it still doesn't explain why they go at all.

During the rainy season, in the days running up to the new moon, thousands upon thousands of female olive ridley sea turtles *Lepidochelys olivacea* arrive at the same time off the beach at Ostional on the Pacific coast of Costa Rica. The raft of bobbing heads in the 'flotilla', as it's known, increases rapidly, and at some unknown signal the turtles begin to come ashore to nest in the black volcanic sands. Local people call it the *arribada*, meaning 'the arrival', and it is the most extraordinary sight – a steady stream of females emerging for several days. The largest known *arribada* was in November 1995, when an estimated 500,000 turtles came ashore.

They start to emerge soon after sunset and continue until early the next morning, using the dark of the night to mask their digging and egg laying from predators, and they deposit so many eggs in the sand – up to 10 million – that any nest raiders are quickly sated so a good number of their eggs will develop and hatch. Even so, turtles in the first waves have little success for the turtles that follow dig up their nests and their eggs in the frantic need to deposit their own. Black vultures *Coragyps atratus*, collared peccaries *Tayassu tajacu*, and white-nosed coatis *Nasua narica*, along with many other egg thieves, are there to pick up the spoils.

About forty-five to fifty-five days later the tiny turtle hatchlings pop up from their nests in the sand and embark on their dangerous journey to the sea. The vultures are there waiting for them, as are the coatis, and they are joined by lizards that scrunch the hatchlings in powerful jaws, ghost crabs *Ocypode* that haul the baby turtles into their burrows, tearing them literally limb from limb, and frigate birds *Fregata magnificens* that swoop in and pluck any daytime stragglers from off the beach. And when they reach the sea, the hazards continue – sharks patrol offshore. It's tough being a baby turtle, but if a young female is lucky and lives to become an adult she will return to the exact same spot to join in future *arribadas*.

The world's biggest sea turtle is the leatherback *Dermochelys coriacea*, the largest known specimen being over 3m (9.8ft) long. It was found washed up dead on a Harlech beach in north-west Wales and you can see it stuffed and mounted in the National Museum of Wales in Cardiff. Apart from its size, the leatherback is unusual among turtles in that it has a rubbery back rather than a shell-like carapace so characteristic of other sea turtles, and its outline, seen from above, is like a teardrop, a good hydrodynamic shape that enables it to power through the water propelled by large front flippers that can be up to 2.7m (8.9ft) long, the largest flippers of any species of sea turtles.

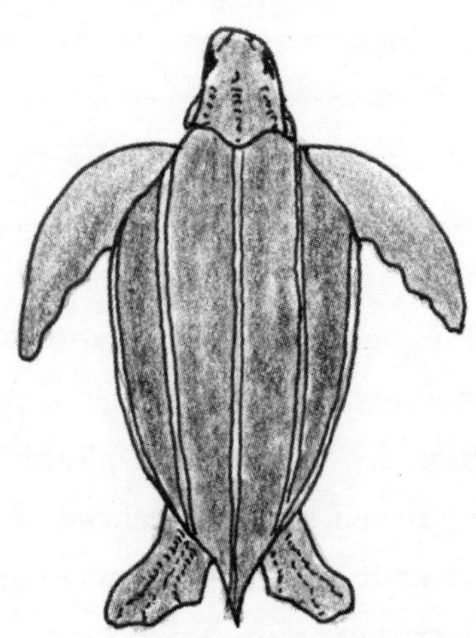

Leatherback turtle

It also has several adaptations for swimming in the cold depths, for this species dives deep to avoid predators at the surface. Counter-current heat exchangers in the blood system, in which heat generated by the muscles is not lost but retained, together with plenty of insulating fatty tissues ensure that the body is warmer than the surrounding seawater. Nearly a quarter of the head and neck, for example, is fatty adipose tissue, with the brain, eyes, oesophagus, salt gland, large arteries and veins wrapped in blubber to insulate against heat loss. With these adaptations, the turtle's body temperature remains at about 25°C (77°F), even though the water temperature at depth can be as low as 0.4°C (33°F).

The leatherback is the most active of turtles, the muscles generating useful heat, so it spends little more than 0.1 per cent of its life resting. It travels great distances across the oceans from breeding sites in the tropics to feeding grounds in cold temperate seas where the turtles feast on jellyfish. It's thought that many sea serpent sightings off the west coast of the British Isles are actually leatherback turtles following the swarms of jellyfish that gather in summer. It's also the deepest diver among turtles, with one individual diving to 1,280m (4,200ft). Dives generally last no more than ten minutes but the extreme divers can be down for as long as seventy minutes.

MISCELLANY OF MARITIME MYSTERIES

The sea is ripe for mystery; after all, it covers three-quarters of the planet and yet we have barely explored it. There is no doubt that there are myriad species of marine animals still to be discovered, and as we have seen in these pages new species are discovered, analysed and classified almost daily. Most are small, such as the purple-coloured, deep-sea acorn worms discovered by University of Aberdeen researchers 2,500m (8,202ft) down on the Mid-Atlantic Ridge and named *Yoda purpurata* in honour of the *Star Wars* diminutive Jedi master Yoda, or the 'giant' virus *Megavirus chilensis* found in the sea near Las Cruces on Chilean coast, which is the largest virus ever found, bigger than many bacteria and visible with an ordinary light microscope. Some discoveries, however, are more like 'giants', such as Perrin's beaked whale *Mesoplodon perrini*, two of which stranded on a beach in California in 1976, formally described in 2002, or the colossal squid *Mesonychoteuthis hamiltoni* (see page 188), which was known only from tentacles found in a sperm whale stomach in 1925 until an intact specimen was caught in 1981, and is thought to grow to 14m (46ft) long. So, why could there not be other large unknown animals out there waiting to be found? On deserted shores, in sheltered coves, from windy headlands or ships at sea, we occasionally get tantalising glimpses of creatures for which we have no identity and which sometimes seem not to adhere to the known laws of nature. Take, for instance, the puzzling case of the truly extraordinary 'sea monkey'.

Sea monkey

In 1965, the distinguished British yachtsman Miles Smeeton was sailing aboard his ketch *Tzu Hang* close to the Aleutian Islands in the North Pacific. He was off the north coast of Atka when his daughter Clio spotted an animal floating in the sea about three metres off the port bow. Smeeton reckoned it was about the size of a sheep, with long, reddish-yellow, pepper-and-salt hair that floated about its body like seaweed on a rock. It was floating on the surface, but as the *Tzu Hang*'s bow approached, it dived beneath the boat. Clio described the head as being like that of a dog rather than a seal, with eyes set close together, again unlike a seal or sea lion. Family friend Henry Combe was also on board and he described it as having a 'face like a Tibetan terrier, with drooping Chinese whiskers'. After it dived, it was not seen again. The description fits superficially the northern fur seal *Callorhinus ursinus*, but the astonished onlookers ruled out a seal or sea lion by way of an explanation, as they had seen many on their voyage;

similarly sea otters were familiar to them, but it looked nothing like any of them. As Smeeton wrote later in *The Misty Islands*: 'In all the thousands of miles that *Tzu Hang* has sailed we have seen nothing that could be described as mysterious or unaccountable, except this one beast...'

And there the matter would have rested – an unexplained phenomenon – had it not been for a similar encounter more than 200 years earlier in the same part of the ocean. In 1741, the naturalist Georg Wilhelm Steller was aboard the Russian ship *Saint Peter* captained by Vitus Bering. They had set out from Kamchatka in search of North America and were returning home, when close to the Semidi Islands (to the east of Atka), Steller described seeing a strange creature in the sea. It had a dog-like head, and pointed ears and whiskers hung down from the upper and lower lips. Its eyes were large, the body round and thick and tapering towards the tail. It was covered with grey and reddish-coloured hair. The animal jumped gracefully and swam about the ship for about two hours, and occasionally it brought the front third of its body right out of the water. It also shot underwater from one side of the ship to the other, and played with and took bites out of a strand of seaweed, what Steller described as doing 'such juggling tricks that one could not have asked for anything more comical from a monkey'. During the voyage Steller had been watching and describing fur seals, sea lions, sea otters and porpoises so the mystery animal is unlikely to have been one of them, but he knew that something similar had been seen about 180 years previously.

Sea monkey

On this occasion, the animal had been sketched by the Dresden-based physician Johann Kentmann and published in Swiss naturalist Conrad Gessner's *Icones animalium* in 1560. Gessner called it the 'sea ape' or 'sea monkey' and gave it the scientific name *Simia marina*. However, what Smeeton and the crew of *Tzu Hang*, Steller and the crew of the *Saint Peter*, and Kentmann's anonymous observer had seen must remain unidentified and unknown, for nothing of this nature has been seen again, let alone caught ... or has it?

The *New York Times* of 5 April 1885 carried two fish stories (the other is told in Caddy (2) below) which were first published in the *San Francisco Chronicle* on

28 March. The first is of two fishermen, Carl Sevening and John Peat, who were rowing near the North Heads in San Francisco Bay when an animal, with 'a fiercely moustached head of a shape between that of a seal and a sea lion, surveyed the scene, took a dislike to the rowboat, and charged upon it'.

The animal dived under the boat and rose up directly below it, lifting it and its occupants skyward, but failed to capsize it and throw the two men into the water. When it surfaced on the other side of the boat it was met with a barrage of blows from the fishermen's oars, which apparently knocked it out. They tied the creature alongside and began towing it to the shore but it came to and towed the rowboat rapidly for about a quarter of a mile. When it came up for a breath, however, it was dealt a final deadly blow and towed to the bottom of Larkin Street (just to the west of Fisherman's Wharf) where it was put on display until noon the following day. The animal was about 2m (6.6ft) long and weighed 136kg (300lb), had green eyes, 'a long, white bristling moustache', and a pair of powerful flippers about 0.46m (1.5ft) long. No identity is offered in the article, save to say that 'the bay yielded a sea monster of such strange appearance that the oldest tar on the sea wall has not yet even given it a name'. Had Sevening and Peat caught a sea monkey? We'll never know. It's another of the sea's many enduring mysteries.

Mermen explained

While the sea monkey cannot really be described as a 'sea monster', there are large and sometimes potentially dangerous sea creatures of one sort or another that have been recorded numerous times down the ages. Just as any large lake in the world has its own monster, so there seems a need for folk living on the coast to believe in some paranormal being that will rise up out of the ocean and terrorise the community. When gazing out to sea, even the most humdrum of creatures can take on a supernatural air, and the physics of the water-air interface has a curious way of distorting what you see. Take the whale and dolphin behaviour known as spyhopping, for example, the way in which sea creatures poke their heads above the surface and take a look round.

When at the Institute of Oceanographic Sciences, Dr Martin Angel once told me about his first 'sea monster' sighting – minke whales spyhopping. The pale countershading of the underside of the head and throat made it look as if the top of the head was on a slender 'neck'. At the University of Manitoba, Dr Waldemar Lehn and his colleague Irmgard Schroeder took the spyhopping explanation for sea monsters a little further. They had read in the thirteenth-century manuscript *The King's Mirror* (see page 183) about a creature called a 'merman' – a tall monster that 'rises straight out of the water ... with shoulders like a man's but no hands ... the body grows narrower from the shoulders down … no one has observed it closely enough to determine whether its body has scales like a fish or skin like man … Whenever the monster has shown itself, men have always been sure a storm would follow.'

Lehn and Schroeder were intrigued, and with a bit of digging came across a similar story in the text *Historia Norvegiæ* written in 1170. They were struck again by the references to storms. So, they put all the available information about atmospheric conditions before a storm into a computer and ran a program that duplicated pre-storm weather images. They used pictures of a killer whale and a walrus spyhopping as their subjects and came up with interesting results. Under certain conditions both animals viewed from about a kilometre away appeared much taller than they really were, with a narrow 'neck' or 'waist' at the base. The killer whale was transformed into a credible sea monster and the walrus into a 'merman'.

The cause of the distortion is a temperature inversion that occurs when a mass of warm air moves over cold air. When light passes through the disturbed air, it is distorted in the vertical plane. In the experimental program Lehn found that a temperature difference of about 7.5°C (13.5°F) with the boundary (thermocline) about 2.2m (7.2ft) above the sea's surface was sufficient to distort the image of, say, the killer whale. In those conditions someone with their eyes a couple of metres above the sea, about the right height for a sailor standing on the deck of a Norse long-ship, would not recognise the whale as a whale but would more likely see a strange monster.

Interestingly, these conditions occur naturally during the last stages of a passing warm front, usually a calm period before a storm, so the ancient Norse mariners had not made up their sea monster stories; they had recorded exactly what they saw. It was their interpretation that was flawed.

So, bearing in mind that sightings of strange creatures could have more rational explanations, like those in *The King's Mirror*, let's take a look at a few eyewitness accounts, for in just a few cases observers really have seen something unusual, something that science cannot readily explain, and these kinds of observations seem all the more believable and possibly more reliable when made by alert, trained eyes. Here are just three unusual sightings.

On the high seas

The year was 1937. Twenty-four-year-old male nurse Alfred Peterson was on board a troopship racing across the South China Sea to relieve the British Concession in Shanghai at the start of the Second Sino-Japanese War.

He was jogging around the deck before breakfast but was stopped in his tracks by something he saw at the sea's surface.

> I looked out and I saw what I thought was a very big tree floating in the sea. A bit later, I stopped on the deck again to get a few minutes' breath and noticed the thing was still level with the ship, and I thought to myself, 'That's not a tree, that's moving with us; a tree couldn't be moving like that.' And so I watched it. I could see the body was about 24 to 25-foot-long, with a tail at the end, all on top of the water. It was grey-black, hippo-coloured. It sort of half-turned, and its neck stuck up like a giraffe's. On top was this giraffe head sort of shape, and I could see two sort of ears or horns or whatever; they were like drumstick ears. It simply played as it went along; it disappeared, came up again, played, and went on. It didn't look vicious; it wasn't splashing; and it looked a big, really gentle thing.

The second example was a year later in 1938. Captain Kingston Lewis was at the time one of the crew aboard the tanker *British Power* on passage in the Persian Gulf between Durban and Abadan. It was a hot day and the sea was calm. Captain Lewis was taking a siesta on deck when his attention was attracted by something in the sea.

> I happened to stand up and I was looking over the side and this animal or monster popped its head out of the water. It had a longish neck about six to eight feet long with a head on it something like a horse. It looked at the ship as if to say, 'What is this?' and after twenty or thirty seconds it ducked its head down into the water and disappeared. I told the second officer about this and he

> sort of laughed it off, he didn't believe me, but not long afterwards he came down and said to me, 'You know, Lewis, you were right about that.' He said, 'It's here in the Persian Gulf Pilot [the sailing directions for the Persian Gulf] and apparently a similar animal had been seen by one of His Majesty's survey ships, and they had sighted a sea animal or monster and made a sketch of it.'

The third sighting took place during the Second World War. In 1942, Mr Welch was on board a troopship bound for Bombay from Durban. He was on lookout duty.

> We never knew quite what we were looking for but we were always on the look-out. On one occasion, though, I could see a large black object way in the distance. My heart went down to my boots because I thought it was a submarine. I sounded the alarm, bells rang all over the ship, and everybody was going mad, panicking. One of the duty officers looked through his binoculars and said, 'Oh no, it's not a submarine, I don't know what it is, probably just something floating in the water.' Anyway, as the ship got nearer we could see what I can only describe as a sea monster; it was definitely something swimming. It crossed our bows and we could see it quite clearly; it was sort of a serpent, about 20 to 30 feet long, very thick – probably as thick as a tree-trunk – and its back was arched in several places. I couldn't make out its head with any clarity; it was sort of surrounded by waves. We carried on, and it went its way as we went ours. It took no notice of us whatsoever, and eventually disappeared from view.

North Atlantic sighting

After the Second World War, the so-called 'Cold War' meant that there were still battalions of observers scanning the oceans, especially aerial surveys to locate potentially hostile submarines. In 1957, just such a mission was about 800km (497mi) north of the Canary Islands in the Atlantic Ocean. The aircraft was a Shackleton from 228 Squadron of the RAF, flying out of St Eval in Cornwall, and the crew were observing a sector of the ocean that was generally devoid of shipping, let alone submarines. Suddenly, they saw a dark shape just below the surface, the telltale sign of a submarine ... or often as not a shark or a whale. They began to circle and drop down to investigate. Then, they saw something quite remarkable and every crewmember on board saw the same thing: the head and neck of a large sea creature that they all agreed looked like a plesiosaur. The skipper told them not to mention the observation, lest they be accused of being drunk on duty, and they all kept their word until recently when the navigator on board told his son, who in turn gave the story to cryptozoology researcher Neil Arnold, co-author of *The Mystery Animals of the British Isles*. It was subsequently published in the *Fortean Times* by Karl Shuker, author of *The Encyclopedia of New and Rediscovered Animals*.

But, what had the RAF crew seen? Could they really have spotted a living plesiosaur? Science of course says no, for no reliable scientific evidence exists for these animals having lived past the end of the Cretaceous period about 65 million years ago when many different species disappeared during a mass extinction event, including the non-avian dinosaurs and their relatives, yet there have been sightings like these, describing remarkably similar animals – serpent-like or plesiosaur-like creatures – right up until the present day.

Newfoundland mysteries

In the early 1990s, Newfoundland became the focus for sea monster hunters when fishermen at Bay L'Argent, 267km (166mi) by road west of St John's on the shore of Fortune Bay, spotted a creature that they described as a 'dinosaur'. In another incident in May 1997, fisherman Charles Bungay encountered a creature with a long neck and grey, scaly skin at Little Bay East, a few kilometres to the north-east of Bay L'Argent. At first he and his companion spotted what they thought were garbage bags and went to investigate. As they came close, however, the creature reared up its head. It had a neck about 2m (6.6ft) long, a head like a horse, with either horns or long ears, and dark eyes. Its total length was estimated to be 9–12m (30–40ft). In a third sighting in April 2000, Bob Crewe was driving along the Cape Shore Road, on the Bonavista Peninsula, when he saw something in the ocean – what looked like a rock where there should not have been one. He stopped his truck, burped his horn and a head popped up. He watched a creature with a long neck and snake-like head swim at speed towards the lighthouse at the northern tip of the peninsula. And more recently, in March 2010, *The*

Telegram published an article that told of the exploits of fisherman John Marsh.

On a summer's day in 2009, Marsh received a call from his son and nephew asking him to come and help for they had snagged a large creature in their capelin traps at Lower Lance Cove off Trinity Bay on the north side of Random Island. He immediately thought that they had caught a whale or porpoise, but when he arrived he saw immediately that the animal had no blowhole like a whale. Its skin was smooth, with no barnacles or scratches, and a green-blue colour. Its teeth were rounded and its lips were camel-like, and it had a three-pointed tail. Marsh told *The Telegram* that he had to leave his son to go to a doctor's appointment but when he returned to the carcass he found that it had sunk out of sight. Unfortunately, there are no photographs, no tissue samples and no carcass (isn't that always the way?) so local experts understandably suggested that Marsh and his son had caught a whale or shark or something similar that had decomposed to the shape he described. A few years previously, for example, the body of a dead creature washed up on the shore on the south coast of Newfoundland. It looked like nothing anybody had seen before, yet when DNA analysis was carried out on its tissues it was positively identified as a sperm whale.

A similar event occurred in Chile in July 2003 when a huge mass of tissue, which became known worldwide as the 'Chilean Blob', was washed ashore at Pinuno Beach in Los Muermos. Speculation was that it was a gigantic deep-sea octopus, but when samples were DNA tested, it was found that the blob was a great mass of adipose tissue from a dead sperm whale. Previously, other sea monster 'blobs', such as the so-called giant octopus of St Augustine in Florida, the Tasmanian West Coast Monster, two Bermuda blobs and the Nantucket Blob, were all found to be decomposing whale tissues.

Even so, sea monsters are a feature in Newfoundland and nearby waters. There is even a record from HMS *Squirrel* dated 1583. Sir Humphrey Gilbert (Sir Walter Raleigh's half-brother) was aboard, having just seized Newfoundland for the British crown. Not long after losing one of his flotilla on the sandbanks of Sable Island, off the coast of Nova Scotia, the *Squirrel* encountered a sea monster which was said to have resembled a lion with glaring eyes.

South Devon predator

Sea monster sightings seem to have been a feature of the seascape since humankind went down to the sea in boats, and they just keep coming. One that caught my eye was on the other side of the Atlantic, at Saltern Cove, near

Goodrington in South Devon. On 27 July 2010, several local people watched an unidentified creature chasing a shoal of fish about 20m (66ft) from the shore. It was taken for a leatherback turtle *Dermochelys coriacea* at first for they are often seen off the coasts of the British Isles, but leatherbacks do not tend to chase fish; they feed on jellyfish. Other sea turtles, such as loggerheads *Caretta caretta*, do feed on fish, but on average they are much smaller than leatherbacks, although 2.79m (9ft) specimens have been claimed and the British Isles are well within the species' range. In fact, loggerheads have been found as far north as the west coast of Scotland. According to eyewitness statements, some of the fish at Goodrington beached themselves in their efforts to escape the unidentified animal. It was 'as long as a sea lion', according to reports received by a spokesman from the Marine Conservation Society, with a long, thin neck about 0.6m (2ft) long and a reptilian head which 'craned above the surface'.

It is quite possible that many sea monster sightings along the west coast of the British Isles are of sea turtles. In September 1959, for example, a well-documented observation took place near Soay in the Western Isles of Scotland. On the morning of the 13th, basking shark fisherman Tex Geddes, who once crewed for Scottish naturalist Gavin Maxwell, and engineer James Gavin, who was on holiday on the island, were out mackerel fishing in a small dinghy. Visibility was excellent. They had seen a basking shark and a pod of killer whales in the distance, but their attention was drawn to a black shape about 3km (2mi) away across Soay Sound, between Soay and Skye. It approached at about 3–4 knots (4–5mph), and as it came closer they could hear it breathing.

In a letter to the naturalist Dr Maurice Burton, who had written extensively on lake and sea monsters, Geddes described the encounter and it was published in the *Illustrated London News* on 4 June 1960. The head was described as 'reptilian' on the top of a 0.76m (2.5ft) long neck. It had large 'protruding eyes', a 'large red gash of a mouth ... which appeared to have distinct lips' and 'no visible nasal organs'. The creature's back 'rose sharply to its highest point some three to four feet out of the water and fell away gradually towards the after end', and the two eyewitnesses 'saw 8 to 10 feet of back on the waterline'.

The creature came to within 18m (59ft) of the boat and constantly turned its head from side to side. The two petrified occupants noticed that the head was blunt, there were no teeth, the body was scaly, and the midline of the back came to a knife-edge ridge, which was deeply serrated. It seemed to breathe through its mouth, which it opened and closed regularly. Geddes estimated that it weighed about 5 tons.

Dr Burton also received a letter from James Gavin. He described the body as being '6 to 8 feet long' at the waterline, about '2 feet high', with the 'line of the back formed by a series of triangular shaped spines, the largest at the apex and reducing in size to the waterline'. He pointed out that the 'neck appeared to be cylindrical' and '8 inches in diameter', and it 'arose in the water about 12 inches forward of the body'. He could see about 15–18 inches of neck, and described the head 'like that of a tortoise' and as 'big as the head of a donkey'. The eye was round, like that of a cow. As well as its appearance, Gavin also described its behaviour at the surface. He wrote:

> At intervals the head and neck went forward and submerged. They would then re-emerge, the large gaping mouth would open (giving the impression of a large melon with a quarter removed) and there would be a series of very loud roaring whistling noises as it breathed. After about five minutes, the beast submerged with a forward diving motion ... It later resurfaced about a quarter of a mile further out to sea and I then watched it until it disappeared in the distance.

Gavin noted that lobster fishermen, fishing north of Mallaig (on the mainland south-east of Soay), had also seen the animal, 'much to their consternation', he added.

Dr Burton looked at the reports and sketches and thought there was an uncanny resemblance to a sea turtle; in fact, bearing in mind that with no reference points, size estimates are difficult at sea, both descriptions strongly suggest a large loggerhead turtle ... or was it?

In the 1970s, Barmouth in North Wales was put firmly on the sea monster map. First, a couple walking on the beach at Llanaber came across large footprints in the sand at the water's edge. Four years later, a group of schoolgirls disturbed a large dark creature, said to be about 3m (10ft) long with feet 'like huge saucers with three long pointed protruding nails'. Their schoolteacher made a sketch of it. More reports followed. On several occasions, the staff at the Minydon Hotel near Red Wharf Bay, Anglesey, saw a strange creature enter the bay. A fishing boat near Bardsey Island at the tip of the Lleyn Peninsula encountered a creature with a long neck and huge body that surfaced next to the boat, and when the fishermen were shown the schoolteacher's sketch they immediately recognised the creature.

Another encounter was from a private yacht about 8km (5mi) from Shell Island (Mochras) near Harlech. It was a sunny afternoon and a calm sea when the husband-and-wife crew spotted what they thought was a seal playing with a couple of tyres. As they got nearer they could see that it was no such thing.

> As we drew closer we thought it was a huge turtle, but it turned out to be unlike anything we'd ever seen. It had a free-moving neck, fairly short, rather like a turtle's, and

> [an] egg-shaped head about the size of a seal's. It's back had two spines, which were sharply ridged, and it was about 8 feet across and 11 feet long, although the ripples in the water when it dived indicated that it was probably twice that length.

Does 'loggerhead turtle' come to mind? Could it be that all these people saw an unusually large loggerhead that wandered into British waters? Without clear photographs we'll never know, and we could fill an entire book with sightings around the world, but before we go, there are a series of events on the north-west coast of North America that are worth a mention. They even prompted two scientists to urge the scientific community to recognise a creature without a proper description and give it a scientific name.

Caddy (1)

In 2009, a couple of salmon fishermen spotted and took a video of a small group of mysterious creatures swimming at the surface in Nushagak Bay, a large estuary that opens into Bristol Bay, the easternmost arm of the Bering Sea, off the coast of Alaska. The wobbly footage taken by Kelly Nash during a rainstorm shows traditional Loch Ness Monster-type humps ahead of which appears to be a longish neck with a small head, but as the neck does not rise clear of the water it could simply be the front part of a large body, most of which is obscured by the water. The animals rise at intervals seemingly to breathe, and at one point a spout of water comes from behind or beside the head of the leading individual.

Professor Paul Leblond, former head of the Department of Earth and Ocean Sciences at the University of British Columbia and a noted cryptozoologist, viewed the video for the Discovery Channel, and suggested the animals captured on tape looked like a crypto-creature known as *Cadborosaurus* – a plesiosaur-like animal that was first spotted in Cadboro Bay in British Columbia and which has been seen many times along the Pacific coast of North America from California to Alaska. So convinced are some researchers that this creature exists they have suggested a scientific name – *Cadborosaurus willsi*, in honour of Archie Wills, a local newspaper editor who, over the years, carried many Caddy stories.

Cadborosaurus is reputed to have a horse-like head, a long neck and large front and rear flippers, and according to Leblond and Edward Bousfield, a former research associate at the Royal Ontario Museum, in their book *Cadborosaurus: Survivor from the Deep*, it is not a new phenomenon. It is thought to arrive on the Pacific Northwest coast every summer, when the waters warm, and it has being doing so for hundreds of years. The Inuit of Alaska even painted a picture of it on their canoes to keep it from attacking, and it is known by a variety of names in many native languages.

'Caddy', as it is known popularly nowadays, came to prominence in the wider world back in 1937, when the partly digested carcass and bones, including a camel-like skull, serpentine skeleton and two-lobed tail with a filamentous fringe, of a 3m (10ft) long unknown animal were taken from the stomach of a sperm whale. The whale was hauled out at the Naden Harbour whaling station, on the north coast of Graham Island in the Queen Charlotte Islands. Samples of the mystery beast were sent to the Nanaimo fisheries station and to the museum in Victoria. The Nanaimo samples went missing (of course) and the samples sent to the museum were identified by curator Francis Kermode as the remains of the foetus of a baleen whale, presumably one that had been aborted, and then they too were lost (surprise, surprise!). James Wakelen, who was one of the flensers at Naden Harbour, believed that Kermode came to a wrong ID. He had seen whale foetuses during the course of his work and strongly disagreed that the creature was a whale foetus, but was at a loss as to say what it was, except that everyone at the station considered it was unusual enough to have it analysed.

However, even though the samples were gone, Captain William Hagelund came across photographs of the creature's remains, which had been laid out along a wall, and published them in his book *Whalers No More* (1987). They show the bones of what is clearly a vertebrate – which could be almost anything really, although the camel-like skull is not that of any known whale or dolphin for it does not have the elongated jaws and rostrum normally seen on these animals. Of course, they could have been damaged during the eating or digesting process, but the skull in the pictures shows a remarkable similarity to the skull of something as equally interesting as a supposed plesiosaur. The Naden Harbour skull closely resembles that of a Steller's sea cow *Hydrodamalis gigas stelleri*, skeletons of which reside in London's Natural History Museum and Washington's Smithsonian.

Steller's sea cow skull with Naden Harbour carcass

In fact, the species is mainly known from skeletons in museums, the German naturalist Georg Wilhelm Steller (1709–46) having been the only scientist to see the creature alive. He wrote:

> The largest of these animals are four to five and three-and-a-half fathoms thick around the region of the navel where they are the thickest. Down to the navel it is comparable to a land animal; from there to the tail, to a fish.
>
> The head of the skeleton is not the least distinguishable from the head of a horse, but when it is still covered

> with skin and flesh, it somewhat resembles the buffalo's head, especially as concerns the lips.

Steller was on Bering's Great Northern Expedition of 1733–43 when he discovered this North Pacific cold-water species of sea cow. He and the crew were marooned for an entire winter on what is now called Bering Island, one of the Commander Islands, about 175km (109mi) east of the coast of Kamchatka, so he had ample time to observe these enormous beasts. He described how these sea cows travel in groups and feed on seaweed.

> The back and half the belly are constantly seen outside the water and they munch along much like land animals ... With their feet they scrape the seaweed from the rocks, and they masticate incessantly ... During eating they move the head and neck like an ox; when a few minutes have elapsed, they heave their head out of the water and draw in fresh air by clearing their throats like horse.

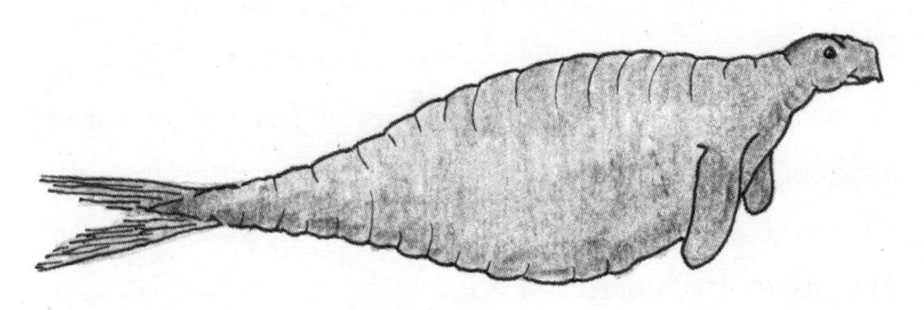

Steller's sea cow

Steller also points out that the flesh of adults is like beef and that of young sea cows tastes like veal, and it was this unfortunate statement that sealed the animal's fate for the fur trappers and traders that followed Bering's ill-fated voyage successfully annihilated the entire Commander Islands population; in fact, Steller's sea cow was thought to have been hunted to extinction by 1768, just twenty-seven years after its first discovery. But, this was not the end of the story: there have been several sightings since.

In July 1962, the Russian exploratory whale catcher *Buran* was close to Cape Navarin on the south side of the Gulf of Anadyr in the north-western sector of the Bering Sea. It was early morning and the ship was close to shore when several of the crew noticed a group of half a dozen large but unidentified animals about 80–100m (263–330ft) away. The following day the ship returned to the same area, a large shallow lagoon, and this time saw a lone individual. A stream fed the lagoon and large quantities of giant kelp *Macrocystis pyrifera* and other marine algae were growing up from the seabed, ideal and ample food for a group of Steller's sea cows, and during winter the lagoon is known to be free from ice, except for a narrow fringe next to the land, so it was a comfortable year-round habitat.

The observers aboard the ship were familiar with whales, seals and sea lions in Russia's Far East, and they were all agreed that the animals they had been watching were neither cetaceans nor pinnipeds. They were dark-skinned with a large body, a relatively small head and an upper lip that appeared to be split. The tail, significantly, was 'bordered by a filamentous fringe', a feature shown also by the Naden Harbour carcass. All the animals swam in the same direction, and were seen to move slowly, dive briefly and resurface, their bodies rising high out of the water, just as the Bristol Bay animals had done. Lengths varied from 6 to 8m (20–26ft), indicating individuals of various ages.

For an explanation and an identity, researchers at the Pacific Research Institute for Fisheries and Oceanography at Vladivostok rejected familiar animals, such as walrus *Odobenus rosmarus* and seals, because lengths never exceed 4m (13ft); instead, they focused on sea cows (Order: Sirenia). The nearest living animal that might fit the description is the dugong, but this is a tropical and subtropical species and very unlikely to end up in the Arctic. The only other animal that seemed to fit the bill is Steller's sea cow. The researchers were almost apologetic in coming to that conclusion; after all, this species was supposed not to exist any more.

However, a search in the archives by V. A. Grekov, published in 1958, revealed that the experience of the crew of the *Buran* was not an isolated case. He found that, although most people believed that the only population of Steller's sea cow was the one near the Commander Islands and that hunters annihilated it, the species had actually been seen elsewhere. Even Steller mentioned that sea cows were to be found around the Chukchi Peninsula, where the local folk used the skins to construct their boats, as well as close to the Americas. There are records, for example, from the late eighteenth century, of them occurring near the Blizhniy Islands, the westernmost islands in the Aleutian Archipelago, where they were called 'Commander cows', and there are several more sightings of sea cows after the supposed extermination date of 1768.

During a round-the-world voyage of 1803–06, the German artist-naturalist Wilhelm Gottlieb Tilesius von Tilenau (the same Tilesius who constructed the 'Adams mammoth' from bones recovered in Siberia) came across reports of sea cows off California. In 1885, Finnish geologist Adolf Erik Nordenskjöld wrote about a sea cow that had been seen the previous year near Bering Island, and in 1930, the eminent Norwegian oceanographer Harald Ulrik Sverdrup recollected being told by a Russian about the body of a dead sea cow being washed up and examined at Cape Chaplin some time between 1910 and 1917. In 1976, Russian fisherman Ivan Nikiforovich is said to have touched a sea cow just to the south of Cape Navarin. In addition, the Pacific Research Institute has from time to time received reports from fishermen in the North Kuril Islands, volcanic islands between Kamchatka and Japan, and from Chukotka, the most north-eastern extreme of Russia, which includes the Chukchi Peninsula.

Other possible sightings are more recent. In 2006 fishing charter skipper Captain Ron Malast (who writes a column for the *Chinook Observer*) and his brother were trolling for tuna about 64km (40mi) offshore from Ilwaco, on the coast of Washington state, when they received a call from another charter boat captain. He had spotted what he called 'a manatee'. Malast headed over and took his boat close to the animal. It was about 3.7m (12ft) long and bigger than a sea lion and Malast writes that he looked it straight in the eye. It remained on the surface for about two to three minutes, seemed quite unafraid, and then slipped gently down into the deep. When Malast and his brother got home they pulled up pictures of manatees *Trichechus* spp. and sea cows *Dugong dugon* on the home computer and were able to confirm that the creature they had seen was, indeed, a manatee-like animal. However, manatees and dugongs are rarely seen in waters with a temperature of less than 18°C (64°F), and North Pacific waters can be considerably colder. West Indian or Florida manatees *T. manatus* have been seen a number of times as far north as Cape Cod in the North Atlantic in late summer, but these sightings have been very much the exception. Aside from a northern elephant seal *Mirounga angustirostris* or a wayward female walrus as possible explanations, Steller's sea cow is again the only other manatee-like option. The ID came up again in 2010.

On this occasion, on 14 September, another fisherman trawling for tuna between Willapa Bay and Long Beach, Washington state, came across a huge, dark grey animal at the surface. Chuck Crosby mentioned it to his wife and, when they looked it up on the internet, they identified the creature as a sea cow. A drawing uploaded to the Cryptomundo website, however, resembles superficially a northern elephant seal, but Crosby found out about the 2006 sighting in roughly the same waters, and was convinced he had seen the same kind of creature – a sea cow and not an elephant seal.

The previous year, in December 2009, there was an unusual incident that involved Canadian Coastguard Auxiliary Unit 59. On a cold, moonlit night, they were on a training exercise off Denman Island, one of the Gulf Islands in the Strait of Georgia to the north-west of Nanaimo. Using a training dummy, the crew were about to test a new hoist mechanism that lifts people from the water when the vessel's navigator Bonaventure Thorburn spotted a strange creature about 300m (984ft) away. Its large head was above the water and it seemed to be resting. At first they thought it to be a whale, then opted for a Steller's sea lion and finally a seal, but it was at least 6m (20ft) long and was deemed not to be one of these familiar animals. As they watched, the animal submerged slowly, leaving a great ring in the water in which clouds of bubbles rose to the surface. The odd thing was their training dummy had disappeared, and it wasn't found until a week or so later on the beach just south of nearby Rosewall Creek.

So, could there be small, relict populations of Steller's sea cows, rather than plesiosaurs, tantalising scientists and crypto-enthusiasts? The Russian researchers believe there is a case to be answered. People from the fur trade, who were seeking fresh meat for their expeditions, wiped out the Commander Islands population, but some of the places mentioned above have few fur-bearing animals and even fewer people, so it's quite possible sea cows have remained undetected, apart from isolated sightings, for the past 200 years. The Bristol Bay and Ilwaco fishermen and the Denman Island coastguards could well have had the privilege of seeing an animal that was thought to be extinct but is actually very much alive ... and is now probably in need of protection.

Caddy (2)

Some of the sightings and sketches of the type of Caddy that is seen around Vancouver Island in British Columbia depict a very different creature to that seen by those west-coast fishermen. The sea cow-type animal has a short but relatively flexible neck and a huge blubbery body, but these other Caddy sightings are almost always of an animal with a long neck and a smoother, more serpent-like body. The head is variously described as 'like a dog' or 'a giraffe' or 'a camel' or 'serpent-like', often with 'ears or horns on the top'. The neck is usually 0.9–3m (3–10ft) long, sometimes with a mane, and the body that is visible above the surface is often 'coiled like half rubber tyres'. On at least two occasions observers have mentioned red eyes. Sightings, which date back to the late nineteenth century, are surprisingly numerous and sketches by eyewitnesses remarkably similar.

However, one notable observation of this type of Caddy was made in 2003 by an unnamed biologist at Taylor Beach, in the Metchosin District of Victoria at the southern tip of Vancouver Island. It was sent to Caddyscan, the organisation that not only coordinates eyewitness reports, but also sets up remote cameras to capture unusual events along the coast. The eyewitness saw a blackish-brown creature with a seal-like head and a long serpentine body that resembled a row of otters undulating across the surface, except they didn't submerge. He also noticed floppy dorsal fins, and the sketch he made at the time has a striking similarity to another drawing made by two observers in San Francisco Bay in 1985. They were Robert and William Clark, and they told me about the time they were sitting in their car just 17m (56ft) from the sea wall on the waterfront at Marina Park beside the bay. The tide was high, the sky blue and visibility good. William Clark was in the driving seat.

> I was looking directly in front when I noticed four or five seals swimming at a fairly rapid speed, about 150 yards away. They suddenly made an abrupt turn and headed for the sea wall. Two seals were moving extremely fast. After a few seconds I saw a wake slightly to the rear of these two seals and looking closer I could see a large black snake-like animal swimming rapidly after the seals.

William could see that it was definitely snake-shaped or eel-shaped, with a clearly discernible head visible about a foot under the water. Behind the head he saw several humps and the animal appeared to be propelling itself by a vertical undulation through the water.

> A series of four coils was created at the front half of its body and these travelled backwards along the length of the neck where they would meet the middle body. At this point the undulation stopped abruptly and slowly dissipated along the length of the remaining part of the body, which was dragged behind. What made it even more astonishing was that this was all happening at a very high speed.

About 20m (66ft) from the sea wall it corkscrewed around so the brothers were able to get a good look at it. It had large hexagonal scales and looked oily or slimy. The head, neck and dorsal part of its body were a dark brownish-green, while the underbelly was a lightish yellow-green. The head was like that of a snake and a little wider than the neck. The neck behind the head was about 25cm (10in.) in diameter. The main part of the body was about a metre (3.3ft) across and the tail was long, and appeared to flatten out towards the end. The entire animal was estimated to be about 18–23m (60–75ft) long.

It paused for a few seconds as if looking for the seals, and then twisted again so its body was once more out of the water. It was then that the two observers noticed two pairs of translucent fan-like fins at the side of the body, which appeared to act like stabilisers. It was these same fins that the Taylor Beach biologist alluded to in his sketch of Caddy.

William and Robert watched the creature disappear and then phoned the coastguard. Later, before a Notary Public, they wrote down their observations and passed them on to the International Society for Cryptozoology. Since then, they have taken a short video of a group of these creatures in the bay, and expert analysis has not pooh-poohed the video images, deeming them to be of scientific interest. However, there's no getting away from the fact that what they saw defies biological conventions – no known fish or marine reptile flexes its body vertically, and although whales, dolphins and seals flex modestly in the vertical plane for they have an up-and-down movement of the tail flukes, they do not have the flexibility to form coils and humps. Similarly, sea snakes have sufficient flexibility in their vertebrae to have 'tail-up' and 'tail-down' positions when swimming fully submerged, which along with air in the lungs helps with changing depth, but they have conventional lateral undulations, like terrestrial snakes, as their main means of propulsion. The Clarks' animals are a conundrum, in fact an impossibility – but it was not the first time a creature like this had been seen in San Francisco Bay.

According to the *New York Times* of 28 March 1865, J. P. Allen of the Bank of California, along with several other Alameda residents, was on the morning ferry from Alameda (Oakland Inner Harbour) when

> a huge black monster suddenly raised its head and neck from the water to a height of about 10 feet, opened its jaws, displaying a mouth two feet wide filled with rows of sharply pointed teeth, and after taking a curious glance at the passing steamer plunged again into the water, at the same time elevating a sixty-foot tail.

The creature apparently shot off and was not seen again, but something similar appeared many years later; in fact, a couple of years before the Clarks' first encounter. It took place a little further north along the coast.

On 1 November 1983 a road construction crew was about 45m (150ft) above the sea on a cliff-top section of Highway 1 just to the south of Stinson Beach in Marin County, north of San Francisco. Traffic controls at both ends of the road works were in radio contact by walkie-talkies, and the elected spokesperson for the crew, safety engineer Marlene Martin, recalled to me what she and the others had seen.

> The flagman at the other end of the job-site hollered, 'What's in the water?'
>
> We all looked out to sea but could see nothing so the flagman, Matt Ratto, got his binoculars. I finally saw the wake and I said, 'Oh my God, it's coming right at us, real fast.'
>
> There was a large wake on the surface and the creature was submerged about a foot under the water. At the base of the cliff it lay motionless for about five seconds and we could look directly down and see it stretched out. I

> decided it must have been about 100 feet long, and like a big black hose about five feet in diameter. I didn't see the tail.
>
> It then made a U-turn and raced back, like a torpedo, out to sea. All of a sudden, it thrust its head out of the water, its mouth went towards the sky, and it thrashed about.
>
> Then it stopped, coiled itself up into three humps of the body and started again to whip about like an uncontrolled hosepipe. It did not swim sideways like a snake, but up and down.
>
> I had the binoculars and kept focused on the head. It had the appearance of a snake-like dinosaur, making coils and throwing its head about, splashing and opening its mouth. The teeth were peg-like and even – there were no fangs. The head resembled the way people draw dragons except it wasn't so long. It looked gigantic and ferocious.
>
> I did not see any fins or flippers and it bothered me that it could move so fast in that way. It was scientifically impossible for anything to go that fast without them. It was not like a snake going sideways: it went up and down.
>
> It stunned me, never in my wildest dreams could I ever have imagined a thing to be so huge and go so fast. I thought, when I saw it, this is a myth.
>
> There were six of us at this time all looking over the rail in disbelief. I was so glad that everybody saw the same thing.

In fact the construction crew were rooted to the spot. A truck driver, Steve Bjora, estimated that the creature had moved at a speed of 80km/h (50mph). When I asked Marlene if the animal had eyes, she hesitated, and then answered.

> I've never really told anybody this before and I cannot swear to it but the eye that I saw was red, a deep burgundy-ruby colour. I've always hated to say that I saw

> that red eye. When I think about the thing I still see that red colour and what's amazing about it is that I've never seen that particular red on anything before.

(In a sighting in June 1946 at Balmoral Beach in Comox, on the eastern side of Vancouver Island, Mrs Winifred Grist also saw a creature with red eyes.)

At first the road crew were reluctant to talk at all about the creature they had seen, but once the word got out several other people called to say they had had similar experiences. Sculptor and minister Tom D'Onofrio recalled a time when he was at the beach at Bolinas, California, at noon one September day in 1976 when suddenly, about 46m (150ft) from the shore, gambolling on an incoming wave, was what amounted to a huge dragon, possibly 18m (60ft) long.

> The serpent seemed to be playing in the waves, thrashing its tail. We were so overpowered by the sight we were rooted to the spot for about ten minutes.

Similarly, Ruth Aryon from Fairfax was visiting her daughter in Bolinas and saw the creature in 1983, and there have been similar strange sea creatures seen in the estuary of the Columbia River, where it is known as 'Colossal Claude', and on the Washington coast, where it became known as the Wachats Serpent.

While seals and sea lions were evident in the California sightings, and thought to be possible prey, the creature's food habits seem somewhat catholic, with fish on the menu as well as several sightings of the animal taking seagulls and waterfowl, swallowing them whole.

What this creature could be is a complete mystery, for no known animal looks and moves in this way, yet the number and quality of the reports show that there must be something out there that defies rational explanation. However, some might say that we want to believe in such creatures and that we adjust our memory of such events to fit an expectation. It was none other than American writer John Steinbeck (1902–68) who encapsulated our need for believing in mysterious creatures from the sea. On one occasion he had heard news that a dead sea serpent had washed ashore at Moss Landing in Monterey Bay but, like the townsfolk, he was mortified when a local scientist had identified it in a matter-of-fact way as a basking shark. He wrote:

> They so wanted it to be a sea serpent. Even we hoped it would be. When sometime a true sea serpent, complete and undecayed, is found or caught, a shout of triumph will go through the world. 'There you see,' men will say, 'I knew they were there all the time. I just had the feeling they were there.'

ABOUT THE AUTHOR

Michael Bright worked as an executive producer with the BBC's world-renowned Natural History Unit, based in Bristol. He is the author of over 130 books on wildlife, travel and conservation, including *Sharks*, *Wild Caribbean* and the bestselling *Africa*, which accompanied the popular David Attenborough series. He is the recipient of many international radio and television awards, including the prestigious Prix Italia.